JEFF HANSON, ASHLEY BERNAL, and JAMES PITARRESI

Library of Congress Control Number: 2024942367

1 2 3 4 5 LBC 28 27 26 25 24

ISBN   978-1-265-64915-9
MHID   1-265-64915-4

Sponsoring Editor
    Robin Najar

Copy Editor
    Alison Shurtz

Production Supervisor
    Richard C. Ruzycka

Art Director, Cover
    Anthony Landi

Acquisitions Coordinator
    Olivia M. Higgins

Illustrator
    Jacob Hanson

Project Manager
    Patricia Wallenburg, TypeWriting

Composition
    TypeWriting

This "unbook" is dedicated to you, the student, whose desire to learn has driven you to seek other methods and sources of knowledge for your success. Students just like you have made the Jeff Hanson YouTube channel one of the most popular engineering education channels on the internet. Through the overwhelming outpouring of success stories and comments from students around the world, the authors were compelled to do even more to further improve student success. We thank you for all of your support and wish you the highest level of encouragement to reach your dreams and make a difference in the world.

Jeff would like to thank his wife Amy for transcribing his notes into legible text, his son Jacob for his exceptional artwork, and his daughter Madison for her moral support.

Special thanks to Jake Ritchie for helping with the Test Yourself and Practice Exam solutions.

About the Authors

Jeff Hanson is a Lecturer in the Department of Mechanical Engineering at Texas Tech University, Lubbock, Texas. He received his PhD in Systems and Engineering Management from Texas Tech. His popular YouTube channel with instructional engineering course videos has hundreds of thousands of followers worldwide.

Ashley Bernal is an Associate Professor of Mechanical Engineering at Rose-Hulman Institute of Technology, Terre Haute, Indiana. She received her PhD in Mechanical Engineering from Georgia Tech and previously worked at Boeing as a Subsystems Engineer on Joint Unmanned Combat Air Systems.

James Pitarresi is a Distinguished Teaching Professor of Mechanical Engineering and Vice Provost and Executive Director, Center for Learning and Teaching Binghamton University, State University of New York. He received his PhD in Civil Engineering at the University of Buffalo.

Contents

Introduction

Years ago, Jeff Hanson began making instruction videos to help his students understand statics followed by mechanics of materials. These videos grew into two entire YouTube channel courses used by hundreds of thousands of students around the world. This "unbook" is the second of the series dedicated to supporting student learning and is the result of over a combined 50 years that the authors have been teaching this topic and finding the best way to explain concepts. We have extensive experience in seeing commonly made mistakes and the most difficult concepts for students to understand, so we not only cover the concepts but also help you avoid the common problems. We call this an "unbook" because we've created a book that looks like the most organized set of class notes ever written for Mechanics of Materials. The goal of every page is to be user friendly with some space for you to work out the problems.

The Structure of the Book

The book is broken into Levels, and as you gain experience and level up, you will be equipped with the skills you need for more and more complex problems. The levels of the book are ordered as you would typically see them in a Mechanics of Materials class, and for each level you will find:

- A **Calculator Problem**, which is a printed, worked out, step-by-step solution
- A **Video Example Problem** with a completely worked out solution on YouTube
- A **Test Yourself Problem** (with the solution in the back of the book) that will make sure you can solve the problem on your own

You'll also find:

- **Pitfalls and Protips** comments regarding common mistakes made by students along with helpful tips to avoid these mistakes
- **Sample Exams** (two complete sets!): One set has complete video solutions, and the other set has fully worked out solutions in the back of the book

How to Use the Videos

All of the videos that accompany the book can be accessed on the Jeff Hanson Mechanics of Materials channel on YouTube using the following QR code:

as well as this link: **https://youtube.com/playlist?list=PLRqDfxcafc21wlI3E56IkDmRJ-33apMjv&si =rKlrvlhe0RtZj5Ff.**

The proper way to use the video solutions is to not just watch the video through because that will make it look easy. Instead, start the video, press pause, and try to solve the problem on your own. Press play if you get stuck and need help in a tough spot. If you watch Tiger Woods play golf on Sunday, you're not going to be playing in the PGA on Monday; you would have to practice—a lot! The same goes for these videos: watching Jeff solve these problems will not guarantee you will be able to solve them. You have to practice working them yourself.

Video solutions to the first practice exam can also be found on Jeff's channel.

Ultimately, your success in this course depends on the amount of practice you put in. The more problems you work, the more comfortable you'll become with this material and the faster you'll be able to solve the problems. So practice these methods, and in no time, you'll be a pro too!

Level 1
Stress

STRESS

 MECHANICS OF MATERIALS LESSON 1
Intro to Solids, Statics Review Example Problem

What Is Mechanics of Materials?

This course has many common names at different universities around the world, including:

- Mechanics of Materials
- Mechanics II
- Solids
- Strengths
- Strength of Materials
- Deformable Bodies

In Statics, we dealt with rigid bodies where no matter the force or moment, there was no deformation, elongation, bending, compression, or failure.

In Mechanics of Materials, we deal with these deformations caused by the forces we calculated in Statics.

PRO TIP	Every Mechanics of Materials problem begins with a Statics problem, so it is imperative that you review your statics notes! The two topics that are absolutely essential are global reactions and internal forces (*M*, *N*, and *V*).

PITFALL	The main mistake made in mechanics of materials is not paying attention to units. In this course it is imperative that you use units in your work. Make sure everything gets canceled. Mixed units happen often and will trip you up big time! Remember:

giga = 10^9 mega = 10^6 kilo = 10^3

milli = 10^{-3} micro = 10^{-6}

Common conversions are as follows:

$1\,N/m^2 = 1\,Pa$ $1\,N/mm^2 = 1\,MPa$

(In order to remember these common conversions, notice the extra "m" in the unit of area and the extra "M" in MPa [i.e., mm and not just m] so it's MPa rather than simply Pa!)

✓ **TEST YOURSELF 1.1**

SOLUTION
TO TEST
YOURSELF:
Stress

Convert the following quantities using your brain and calculator, not an online reference:

Recall: 1 in = 2.54 cm; 1 lbf = 0.454 kg; 1 rev = 2π radians; 1 psi = 6,895 Pa.

1. 2 ft to mm
2. 170 lbs to kg
3. 150 lb/ft^3 to kg/m^3
4. 32 psi to kPa
5. 2,500 rpm to radians/s

ANSWERS

1. 609.6 mm
2. 77.2 kg
3. 2,405 kg/m^3
4. 220.6 kPa
5. 261.8 radians/s

STRESS

STATICS REVIEW

Constructing Free Body Diagrams

- Determine what is the body of interest.
- Cut the body away from the system and "free" it up.
- Draw all forces acting on that body.

Remember: There are several types of connections that generate reaction forces. To determine what forces and moments to add to the body because of these reactions, simply ask yourself, "Can I move in the x, y, or z direction?" and "Can I rotate around the x, y, or z axes?" If the answer is ever *no* to the "move" question, then there must be a reaction force. If the answer is ever *no* to the "rotate" question, there must be a moment preventing that action. Ropes, cables, chains, or cord are always in tension (i.e., going away from or pulling on the body). You can't push a rope, but you can pull on it!

Types of Connections	Reactions	Description of Reaction(s)
		Roller Connections Only one reaction force. Always normal to the plane of contact.
		Pin Connection Has two reaction forces, an x component and a y component.
		Cantilever Connection or Fixed Connection Has an x component and a y component, as well as a reaction moment in the z direction.
		Single Point of Contact Has a normal reaction force perpendicular to the plane of contact.

STRESS

Types of Connections	Reactions	Description of Reaction(s)
		Simple Cable Connection Always drawn in tension along the line of action of the cable.
		Smooth Pin in a Slot Has single normal force reaction always perpendicular to the slot.
		Smooth Collar Has a single normal reaction force perpendicular to the pipe, as well as a reaction moment.

Writing Equations of Equilibrium

For 2D problems, three equations:

$$\Sigma F_x = 0 \qquad\qquad \Sigma F_y = 0 \qquad\qquad \Sigma M_z = 0$$

For 3D problems, six equations:

$$\Sigma F_x = 0 \qquad\qquad \Sigma M_x = 0$$
$$\Sigma F_y = 0 \qquad\qquad \Sigma M_y = 0$$
$$\Sigma F_z = 0 \qquad\qquad \Sigma M_z = 0$$

STRESS

Converting Distributed Loads into Concentrated Loads

- Calculate the total *magnitude* (i.e., the total force) of a distributed load by computing the "area" under the distributed load. This is the equivalent point load.
- Find where the equivalent point load acts by determining the centroid of the area of the distributed load and apply that point load found at the centroid of the shape. Refer to the table in the back of the book for the centroid location of common shapes.

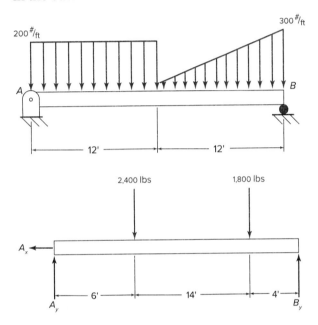

Method of Joints and Method of Sections

Method of Joints

- Find global equilibrium (i.e., find the reactions).
- Select a joint and draw FBD. **Note:** *Do not* pick a joint with more than two unknowns since we only have two equations ($\Sigma F_x = 0$ and $\Sigma F_y = 0$). Since all forces on the FBD of a joint pass through the same point, you can't use the moment equation as there is no distance.

EXAMPLE FBD OF METHOD OF JOINTS

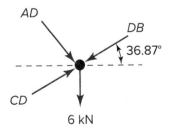

Method of Sections

- Use this method when a problem asks for forces in specific members (not the forces in every single member).
- Find global equilibrium (i.e., find the reactions).
- Cut through the member of interest and draw the easiest side. **Note:** Do *not* cut through more than three members since we only have three equations ($\Sigma F_x = 0$, $\Sigma F_y = 0$, and $\Sigma M_z = 0$) (i.e., this method has FBD where all forces do not pass through the same point unlike the Method of Joints so there is a moment equation).

EXAMPLE FBD OF METHOD OF SECTIONS

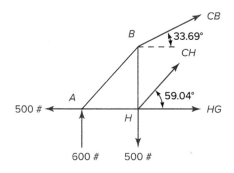

Finding Internal Loads in a Member

Recall: Every time you cut a beam, you must have an M, N, V!

Internal Forces:

- M: bending moment
- N: normal force
- V: shear force

Also remember the positive sign convention for those forces.

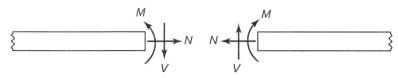

STRESS

EXAMPLE: INTERNAL FORCES

For the given loaded beam, find the internal forces on the beam at point E.

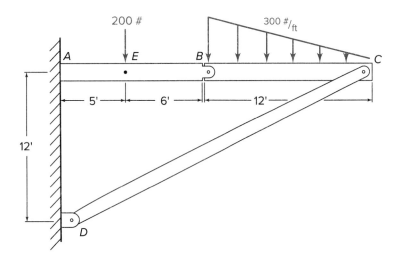

STEP 1: Find global equilibrium. (***Note:*** This is a frame problem from Statics, which has too many unknowns to initially solve.)

- Blow the frame apart and draw the FBD of each part.
- Don't forget when going from one FBD to the next to reverse your assumption for the arrow directions if both diagrams have the same force.

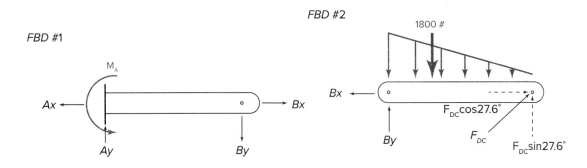

From FBD #2:

$$\Sigma M_B = 0 = F_{DC}(sin(27.6°))(12) - 1{,}800(4)$$
$$F_{DC} = 1{,}295.1 \text{ lbs}$$

Recall: 1,800 lbs is the magnitude of the distributed load and is applied at the load's centroid. Also remember (F_{DC}) member DC is a "two force member," which is pin-connected on both ends with no forces or moments applied to it.

$$\Sigma F_x = 0 = F_{DC}(cos(27.6°)) - B_x$$
$$B_x = 1,295.1(cos(27.6°))$$
$$B_x = 1,147.7 \text{ lbs}$$

$$\Sigma F_y = 0 = F_{DC}(sin(27.6°)) - 1,800 + B_y$$
$$B_y = 1,800 - 1,295.1(sin(27.6°))$$
$$B_y = 1,200.0 \text{ lbs}$$

Now solve FBD #1:

$$\Sigma F_x = 0 = B_x - A_x$$
$$A_x = B_x = 1,147.7 \text{ lbs}$$

$$\Sigma F_y = 0 = A_y - B_y$$
$$A_y = B_y = 1,200.0 \text{ lbs}$$

$$\Sigma M_A = 0 = M_A - B_y(11)$$
$$M_A = 1,200.0(11)$$
$$M_A = 13,200 \text{ lb·ft}$$

STEP 2: Cut through the POI (point of interest), which in this problem is point E.

- Don't forget . . .
  Every time you cut a beam, you must have an M, N, V! ♫♫♪♪♪
- Also recall the positive sign convention for drawing M, N, V.

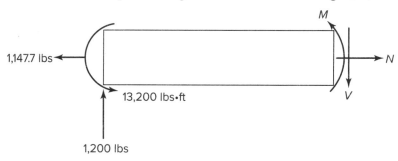

STEP 3: Solve for internal forces M, N, and V.
$$\Sigma F_x = 0 = N - 1,147.7$$
$$N = 1,147.7 \text{ lbs}$$
$$\Sigma F_y = 0 = 1,200 \text{ lbs} - V$$
$$V = 1,200 \text{ lbs}$$
$$\Sigma M_E = 0 = M_E - 1,200(5) + 13,200$$
$$M_E = -7,200 \text{ lb·ft}$$

Note: M_E was assumed to be counterclockwise, but since we computed a negative for M_E it will actually be a clockwise moment.

These are really good "types" of problems to review from Statics because they will be utilized in Mechanics of Materials often.

STRESS

Press pause on video lesson 1 once you get to the workout problem. Only press play if you get stuck.

Find the internal forces at point *B*.

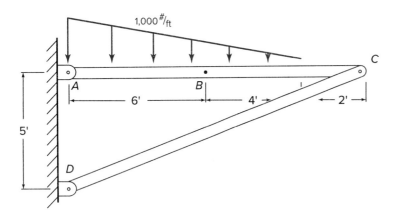

SOLUTION TO TEST YOURSELF: Stress

STRESS

TEST YOURSELF 1.2

For the frame shown, compute:

1. The support reactions at A and B.
2. The Moment M, Shear V, and Normal Force N, in the horizontal member at a location 4 ft to the right of pin A.

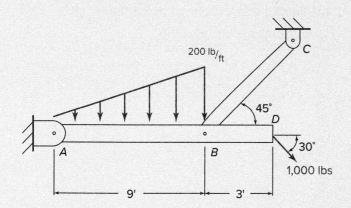

200 lb/ft

45°

30°

1,000 lbs

9′   3′

 **WATCH VIDEO** — **MECHANICS OF MATERIALS LESSON 2**
Normal Stress, Review of Units

What Is Stress?

Stress can be defined as the "intensity of a force."

$$\text{Stress} = \frac{\text{Force}}{\text{Area}}$$

Imagine you can go dancing with Dr. Hanson or the lovely Maggie . . . before you choose, understand that both are really bad dancers, and you are going to get stepped on!

Hanson dances in boots and Maggie in high heels.

Normal Force:

Hanson = 220 lbs

Maggie = 125 lbs

Heel Size:

Hanson = 4" × 4" = 16 in²

Maggie = 0.25" × 0.25" = 0.0625 in²

$$\text{Stress for Hanson} = \frac{220 \text{ lbs}}{16 \text{ in}^2} = 13.75 \text{ psi}$$

$$\text{Stress for Maggie} = \frac{125 \text{ lbs}}{0.0625 \text{ in}^2} = 2{,}000 \text{ psi}$$

As you can see, there would be a much more "intense" force if the smaller weight is applied over a very small area; therefore, it would have a much higher stress.

Normal Stress

- Results from tensile or compressive loading (making things longer or shorter in length, axially).
- Denoted by the Greek letter σ (sigma)
- $\sigma = \dfrac{P}{A}$ also written as $\sigma = \dfrac{N}{A}$
  - P or N is the normal force, either tension or compression.
  - A is the original cross-sectional area of the specimen being stretched or compressed.
  - The sign of stress is as follows:
    - positive for tension $(+)$
    - negative for compression $(-)$

Stress Element

Often in Mechanics of Materials we are interested in the stress at a specific point. We can describe that stress by drawing an infinitesimally small cube at the point of interest (in this case point A). You then add arrows to the stress element to describe the stress state. In this case we can simply draw the stress cube in 2D (xy) since there is no stress in the z direction. If we apply tension to it, the stress element elongates, and if we apply compression it contracts.

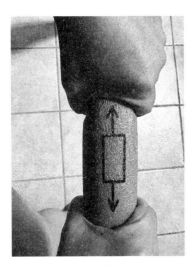
Tension

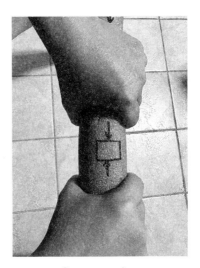
Compression

PRO TIP

An easy way to remember the sign convention for tension or compression is to think of the word "tension" as starting with a (+) positive sign.

The stress on the top and bottom (in the image above) has to be the same magnitude but opposite direction as otherwise the element would "move."

EXAMPLE: NORMAL STRESS

Find the normal stress at point *B*. The cross section as shown is 1.5" by 3".

Cross Section

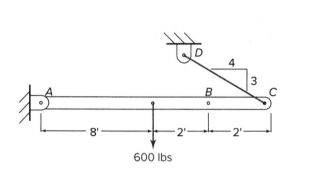

600 lbs

STEP 1: Find global equilibrium.

Ax ⟶

Ay 600 lbs

$$\Sigma M_A = 0 = \frac{3}{5}F_{CD}(12) - 600(8)$$

$$F_{CD}(0.6)(12) = 4{,}800$$

$$F_{CD} = 666.7 \text{ lbs}$$

$$\Sigma F_x = 0 = A_x - \frac{4}{5}F_{CD}$$

$$A_x = 0.8(666.7)$$

$$A_x = 533.3 \text{ lbs}$$

$$\Sigma F_y = 0 = A_y - 600 + \frac{3}{5}F_{CD}$$

$$A_y = 600 - 0.6(666.7)$$

$$A_y = 200 \text{ lbs}$$

STEP 2: Section the beam through the point of interest (point *B*). Remember, you can draw the left or the right side FBD.

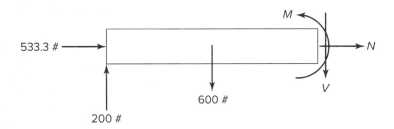

Note: We only need *N* here to solve this problem.

$$\Sigma F_x = 0 = 533.3 + N$$
$$N = -533.3 \text{ lbs}$$

N, here, was assumed to be in tension but it is actually in compression, which means stress at point *B* will be negative.

STEP 3: Calculate stress.

$$\sigma = \frac{N}{A} = -\frac{533.3 \text{ lbs}}{(3)(1.5)\text{in}^2} = -118.5 \frac{\text{lbs}}{\text{in}^2} \text{ or } -118.5 \text{ psi}$$

▶ Press pause on video lesson 2 once you get to the workout problem. Only press play if you get stuck.

If the diameter of the cables is 20 mm, find the normal stress in each cable if the stoplight weighs 100 N.

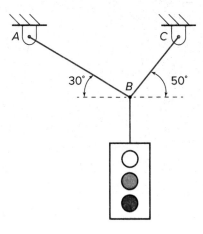

STRESS

✓ TEST YOURSELF 1.3

SOLUTION
TO TEST
YOURSELF:
Stress

Compute the normal stress in members *GF*, *GD*, and *CD* for the truss shown given the cross-sectional areas in the table as *GF* is 5 in², *GD* is 3 in², and *CD* is 6 in².

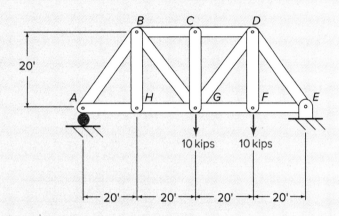

Member Area

Member	Area
GF	5 in²
GD	3 in²
CD	6 in²

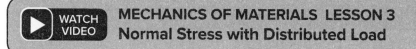
▶ WATCH VIDEO

MECHANICS OF MATERIALS LESSON 3
Normal Stress with Distributed Load

Normal Stress with Distributed Load

You may be familiar with a distributed load perpendicular to the beam (*left image*), but you can also have distributed load along the length of the beam (*right image*) and not just perpendicular.

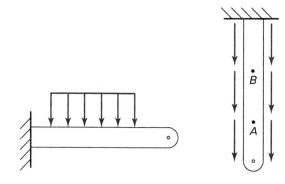

- The image on the right leads directly to a normal stress; the image on the left leads to bending stress (how to calculate bending stress is in Level 6).
- To determine the normal stress at a point, think about the weight you would feel if you were at point A versus point B—point A has lots less force than point B.

EXAMPLE: DISTRIBUTED NORMAL STRESS

A rod of uniform rectangular cross-sectional area A and uniform density ρ hangs downward. Compute the normal stress as a function of the length.

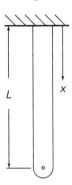

STEP 1: Cut the beam at a distance of "x."

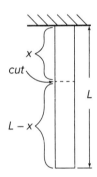

STEP 2: Draw the FBD of the beam of either side (the top or the bottom) (i.e., above or below the cut). Let's draw the FBD of the bottom side.

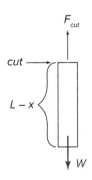

STEP 3: Write the equation(s) of equilibrium (i.e., sum of forces in x or y direction) and solve. In this particular problem you will need to solve for the weight in terms of the volume. (You can do this by multiplying the unit weight ($\rho \times g$) by the volume of the cylinder below the cut, which is the area times the length (which is $L - x$).

$$\Sigma F_x = 0$$
$$-F_{cut} + \rho gA(L - x) = 0$$
$$\therefore F_{cut} = \rho gA(L - x)$$
$$\sigma_n = F_{cut}/A = \rho g(L - x)$$

▶ Press pause on video lesson 3 once you get to the workout problem. Only press play if you get stuck.

Write an equation for the normal stress as a function of x from 0.5 m $< x <$ 1.25 m. The diameter is 100 mm.

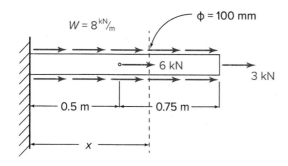

TEST YOURSELF 1.4

SOLUTION
TO TEST
YOURSELF:
Stress

Sometimes when building a foundation for a building, a steel "pile" is driven into the ground and used as a base for the structure. If we assume that the friction of the soil on the pile is proportional to the depth squared ($f = kx^2$), compute the normal stress on a cross section of the pile as a function of depth if the load on the pile is 50 kips. The cross-sectional area of the pile is 10 in^2.

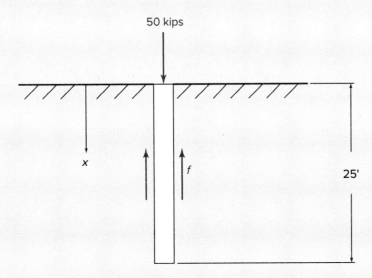

ANSWER

$$\sigma_n = 5{,}000 - 0.32x^3 \ \text{lb/in}^2$$

MECHANICS OF MATERIALS LESSON 4
Shear Stress, Single and Double Shear Example

Shear Stress

 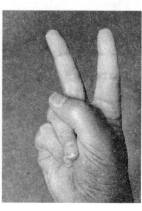

- Results from a perpendicular force that can be thought of as a "tearing force." There will be an upward force on one side of the "tear" and a downward force on the other side.
- If I asked you to go to the kitchen and get shears, what would you come back with?
- How do scissors work? One blade pushes up while the other pushes down . . . precision "tearing" machine.
- What's the international sign for scissors? (Get it, it makes a "V" not a "P"—this should help you remember the equation for shear stress.)
- Shown below are two different types of pin connections. Pin *A* is in what we call double shear and pin *B* is in single shear.

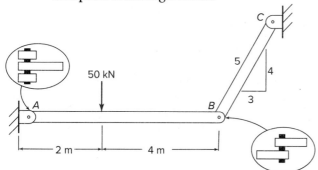

- Denoted by the greek letter τ (tau).

$$\tau = \frac{V}{A} \qquad\qquad \tau = \frac{V}{2A}$$

    Single Shear   Double Shear

- *V* (like the shape of the scissors) is the shear force.
- *A* is the cross-sectional area of the specimen being sheared.

PRO TIP

When calculating shear stress, the cross-sectional area can be confusing to select. Think of shear as "tearing" and think about which area you would have to tear through to make your specimen two separate pieces.

STRESS

✓ TEST YOURSELF 1.5

SOLUTION
TO TEST
YOURSELF:
Stress

Which of these connections are single shear, and which are double shear? Note for the hitch there is one for the ball and one for the pin.

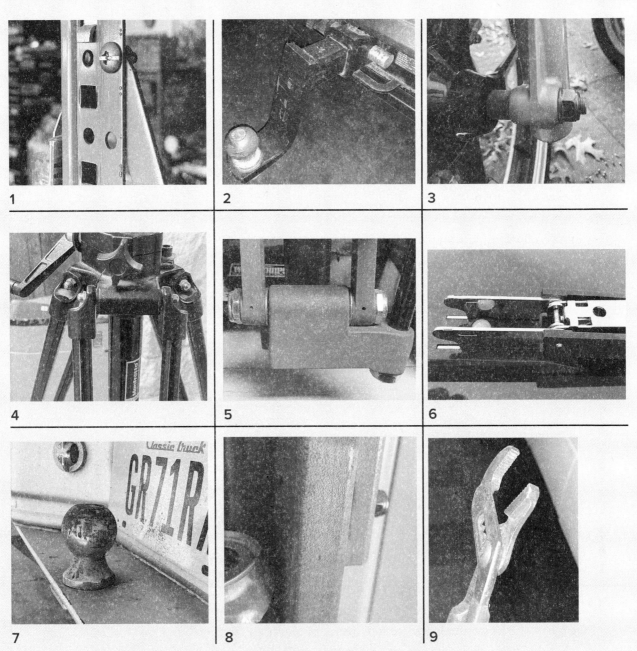

(1) Single, (2) Double, (3) Single, (4) Double, (5) Double, (6) Double, (7) Single, (8) Single, (9) Single

ANSWERS

EXAMPLE: SHEAR STRESS

If max allowable shear stress of the pin at point *A* is 5,000 psi, find the minimum diameter of the pin at point *A* to the nearest $\frac{1}{8}$".

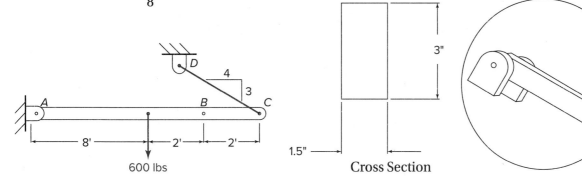

Cross Section

STEP 1: Find global equilibrium. See previous "Example: Normal Stress."

$A_x = 533.3$ lbs (to the right)

$A_y = 200$ lbs (upward)

STEP 2: Find the shear force (*V*) that is tearing the pin at *A*.

Since we need a single force, not an "*x*" and a "*y*" force, find magnitude:

$V = \sqrt{533.3^2 + 200^2} = 569.6$ lbs

STEP 3: Solve for diameter by using the double shear equation.

$$\tau = \frac{V}{2A} \qquad 5{,}000 \text{ psi} = \frac{569.6 \text{ lbs}}{2(\pi r^2)\text{in}^2}$$

$$r^2 = \frac{569.6}{2\pi(5{,}000)}$$

$$r^2 = 0.0181$$

$$r = 0.135"$$

So, $D = 0.135" \times 2 = 0.270"$

Round up to nearest $\frac{1"}{8}$ therefore 0.375 in or $\frac{3"}{8}$ pin diameter.

Press pause on video lesson 4 once you get to the workout problem. Only press play if you get stuck.

Find the average shear stress on the 20 mm diameter pins at *A* and *B* for the given load.

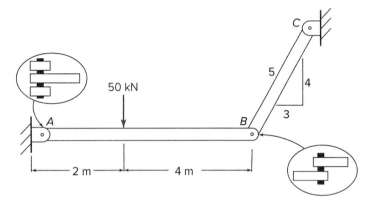

STRESS

TEST YOURSELF 1.6

SOLUTION
TO TEST
YOURSELF:
Stress

For the frame analyzed in Test Yourself 1.2, Pin A is attached with two plates while Pin B uses a single plate connection. If each pin uses a $\dfrac{3"}{8}$ diameter bolt, compute the average shear stress in the pin at A and B.

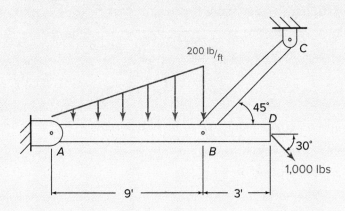

STRESS

▶ WATCH VIDEO **MECHANICS OF MATERIALS LESSON 5**
Bearing Stress Explained, Example Problem

Bearing Stress

What Is Bearing Stress?

Bearing stress is another form of normal stress and therefore is calculated the exact same way as normal stress.

The only difference is that the area for bearing stress is the contact area between the two bodies "bearing" on each other.

Common bearing stress situations:

- A washer under a bolt is distributing load and is experiencing bearing stress. Don't forget that a bolt with a washer is in tension whereas the material being "sandwiched" is in compression.
- A load being carried by two flat surfaces in contact.
  - The total "contact area" is experiencing compressive bearing load.

Why do we even care about bearing stress?

The bearing stress could cause local failures at the site of contact. An example in the image above is that the wood fibers could become crushed at the site of the rectangular plate.

Another example is ensuring that the ground (i.e., soil) can support the intended building's structure. If the soil is too "weak," the building could tilt.

EXAMPLE: BEARING STRESS

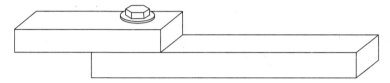

A bolt with a diameter of $\frac{1"}{2}$ is connecting two bars. The bolt is tightened such that a tensile load of 1.5 kips is experienced. If the washer has a 1.75" O.D. and a $\frac{1"}{2}$ I.D., calculate the bearing stress between the washer and the wood.

STEP 1: Calculate the area of the washer.

$$A = \pi(r_o^{\,2} - r_i^2)$$

$$A = \pi\left(\left(\frac{1.75}{2}\right)^2 - \left(\frac{5}{2}\right)^2\right)$$

$$A = \pi(0.875^2 - 0.25^2)$$

$$A = \pi(0.766 - 0.0625)$$

$$A = \pi(0.7035) = 2.21 \text{ in}^2$$

STEP 2: Calculate bearing stress.

$$\sigma = \frac{P}{A}$$

$$\sigma = \frac{-1.5 \text{ kips}}{2.21 \text{ in}^2} = -0.679 \text{ ksi}$$

So, why is it negative? Remember compressive stress is expressed as negative, while tensile stress is positive.

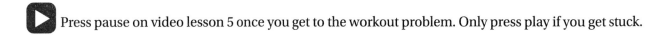

Press pause on video lesson 5 once you get to the workout problem. Only press play if you get stuck.

Bar *BCD* is 15 mm thick, and the bolt at *C* has a diameter of 9 mm and is in single shear. If the force *P* applied is 250 N:

1. Find the average shear stress of the bolt.
2. Find the bearing stress from the bolt exerted onto bar *BCD*.

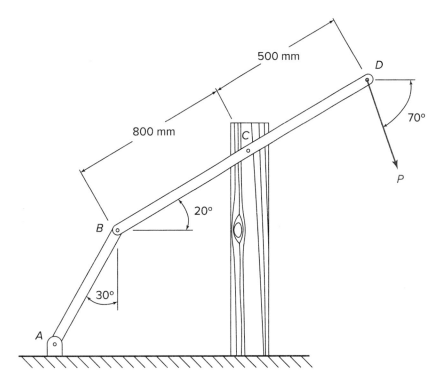

TEST YOURSELF 1.7

SOLUTION
TO TEST
YOURSELF:
Stress

STRESS

For the pinned single shear connection shown below with axial load
$P = 7.5$ kips, determine the bearing stress on the plate due to the bolt.
The bolt diameter is 1.00 in. Next, compute the normal stress on the plate at the location of the hole.
(***Note:*** The hole reduces the cross-sectional area in the plate at that location. This is called the
"net cross-sectional area.")

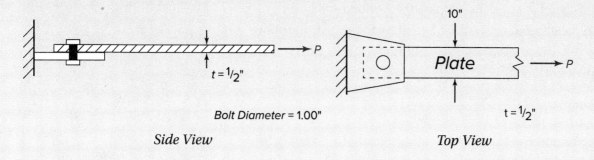

ANSWERS

Normal stress = 1,667 psi

Bearing stress = 15,000 psi

 **MECHANICS OF MATERIALS LESSON 6**
Factor of Safety Explained, Example Problem

Factor of Safety (FoS)

- How much stronger something is compared to its load (example: if a beam can hold 100 lbs and is loaded with 50 lbs, the factor of safety would be 2).
- It is good to know for people that overload stuff.
- Typically, engineers consider failure to be anything above $\sigma_{allowable}$ or what we call σ_Y (yield stress, which will be discussed in detail later).

$$\text{FoS} = \frac{\text{material failure property}}{\text{allowable (your design)}}$$

- Common FoS is 1.5-3 for many applications except in life-harming applications (i.e., cable for emergency evacuation on helicopters) or when there are tremendously too many unknowns regarding the loads expected.
  - A $\frac{1}{2}$ ton truck may have a safety factor of as much as 3.
  - A rocket, which is super concerned about extra weight, may have a safety factor as low as 1.
  - A swingset on a playground near a college campus needs to be designed to support not only toddlers and children swinging but also college students' late night shenanigans.
- A good design engineer *never* forgets to include a factor of safety.

PITFALL

Do I multiply by factor of safety or divide?
This is the most common mistake with these problems. Use your intuition to make the correct choice. Bolts, beams, and cable diameters need to be multiplied by factor of safety to make them bigger or strong than they need to be, whereas something limited by σ_Y or $\sigma_{allowable}$ should be divided by factor of safety to set your design limits lower to make sure to prevent failure.

EXAMPLE: FACTOR OF SAFETY

Using a factor of safety of 1.5, and knowing that the connections at A and E are double shear, find the minimum diameter of the bolts at A and E. $\tau_{allowable}$ for the bolts is 250 MPa.

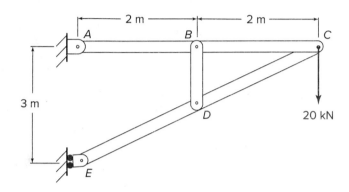

STEP 1: Find the forces at all joints and connections. It is important to realize that this is a frame problem. First, find all possible global reactions.

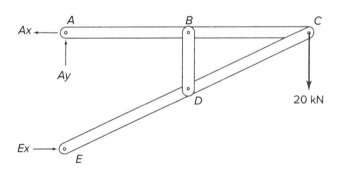

$$\Sigma M_A = 0 = -20(4) + E_x(3)$$

$$E_x = \frac{80}{3} = 26.67 \text{ kN}$$

$$\Sigma F_x = 0 = E_x - A_x$$

$$E_x = A_x$$

$$A_x = 26.67 \text{ kN}$$

$$\Sigma F_y = 0 = -20 + A_y$$

$$A_y = 20 \text{ kN}$$

STEP 2: Blow the frame apart to find all remaining unknowns.
Note: Member *BD* is a two-force member.

FBD #1

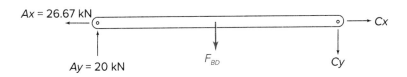

FBD #2

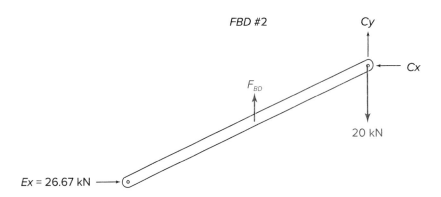

Note: Point *C* is what I call a "yucky" point. Does the 20 kN force go on the first FBD or the second? You can put it on either FBD; however, we can also take the moment about the "yucky" point to avoid dealing with it.

For FBD #1:

$$\Sigma M_C = 0 = F_{BD}(2) - A_Y(4)$$
$$F_{BD} = 2A_Y$$
$$F_{BD} = 2(20) = 40 \text{ kN}$$

For FBD #2:

$$\Sigma M_C = 0 = -F_{BD}(2) + E_Y(3)$$
$$F_{BD}(2) = 26.67(3)$$
$$2F_{BD} = 80$$
$$F_{BD} = 40 \text{ kN}$$

STEP 3: Calculate the total force on pins A and E.

Pin A: $A_x = 26.67$ kN

$A_Y = 20$ kN

total force $= \sqrt{26.67^2 + 20^2} = 33.34$ kN

Pin B: $E_x = 26.67$ kN

total force $= 26.67$ kN

STEP 4: Calculate diameter by using double-shear stress equation.

$$\tau = \frac{V}{2A}$$

$$\frac{\tau}{F.S.} = \frac{33.34 \text{ kN}}{2(\pi r_A^2)}$$

$$\frac{250 \text{ MPa}}{1.5} = \frac{33.34 \text{ kN}}{2(\pi r_A^2)\text{mm}^2}$$

$$166.7 \frac{\text{N}}{\text{mm}^2} = \frac{33,340 \text{ N}}{2(\pi(r_A^2))\text{mm}^2}$$

$$r_A^2 = 31.85 \text{ mm}^2$$

$$r_A = 5.64 \text{ mm}$$

STRESS

Press pause on video lesson 6 once you get to the workout problem. Only press play if you get stuck.

Find the minimum pin diameter at B and C to the nearest $\frac{1"}{4}$ if using a FoS of 1.5 and $\tau_{allowable}$ is 12 ksi.

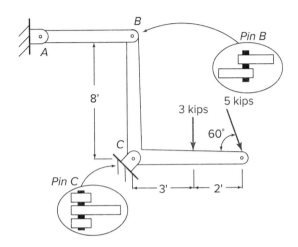

STRESS

✔ TEST YOURSELF 1.8

SOLUTION TO TEST YOURSELF: Stress

For the connection shown, what diameter bolt should be used if the load $P = 7.5$ kips, the factor of safety is 1.75, and the allowable shear stress in the bolts is 15 ksi? Select a bolt diameter to the nearest $\frac{1''}{4}$ and compare the solutions if the connection is in single shear or double shear. The plate has a thickness $t = \frac{1}{2}$ in.

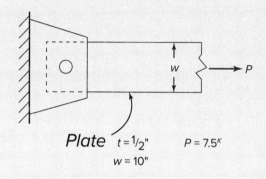

Plate $t = 1/2''$ $P = 7.5^K$
$w = 10''$

Single-Top View

Double-Top View

ANSWER
Single shear use 1.25" bolt
Double shear use 0.75" bolt

Statics

Two most important things to review are:

- Global Reactions
- *M*, *N*, and *V*

Stress: "Intensity of a Force"

There are three kinds of stress:

- **Normal Stress** (σ). Stress caused by either an axial tension or compression (i.e., perpendicular to the cross-section).
- **Shear Stress** (τ). Stress caused by a pair of forces of equal magnitude and opposite direction that are parallel to the cross section. Think about tearing stress. (What area do I have to tear through? It can be either single or double area.)
- **Bearing Stress** (σ). Always compressive and is the result of two areas that are in contact being pushed together. It is another form of normal stress.

Failure

- People will overload stuff, so typically we design with a Factor of Safety (FoS).
- FoS—How many more times safe something is (simply a multiplier used to make something safer than it needs to be).

Equations Learned in this Level

Normal Stress

$$\sigma = \frac{N}{A}$$

Shear Stress

$$\tau = \frac{V}{A} \qquad\qquad \tau = \frac{V}{2A}$$

Single Shear   Double Shear

Bearing Stress

$$\sigma = \frac{N}{A}$$

Factor of Safety

$$\text{FoS} = \frac{\text{material failure property}}{\text{allowable (your design)}}$$

PRO TIPS

Statics Review

- Every Mechanics of Materials problem begins with a Statics problem, so it is imperative that you review your statics notes!

  The two topics that are absolutely essential are global reactions and internal forces (M, N, and V).

Normal Stress

- An easy way to remember the sign convention for tension or compression is to think of the word "tension" as starting with a (+) positive sign.

  The stress on the right side and left side of the stress element has to be the same as otherwise the element would "move."

Shear Stress

- When calculating shear stress, the cross-sectional area can be confusing to select. Think of shear as "tearing" and think about which area you would have to tear through to make your specimen two separate pieces.

PITFALLS

Statics Review

- The main mistake made in mechanics of materials is not paying attention to units. In this course it is imperative that you use units in your work. Make sure everything gets canceled. Mixed units happen often and will trip you up big-time! Remember:

giga = 10^9	mega = 10^6	kilo = 10^3
milli = 10^{-3}	micro = 10^{-6}	

Common conversions are as follows:

1 N/m² = 1 Pa1   N/mm² = 1 MPa

(In order to remember these common conversions, notice the extra "m" in the unit of area and the extra "M" in MPa [i.e., mm and not just m] so it's MPa rather than simply Pa!)

FoS

- Do I multiply by factor of safety or divide?

This is the most common mistake with these problems. Use your intuition to make the correct choice. Bolts, beams, and cable diameters need to be multiplied by factor of safety to make them bigger or strong than they need to be, whereas something limited by σ_Y or $\sigma_{allowable}$ should be divided by factor of safety to set your design limits lower to make sure to prevent failure.

Level 2
Strain

Normal Strain

Shear Strain

STRAIN

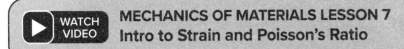

WATCH VIDEO
MECHANICS OF MATERIALS LESSON 7
Intro to Strain and Poisson's Ratio

What Is Strain?

Simply put, strain is deformation. Strain is going to result from one of two loads: normal and shear.

Normal Strain (Axial Load). When a body is loaded axially, no matter the material, it will experience a displacement or a change in length.

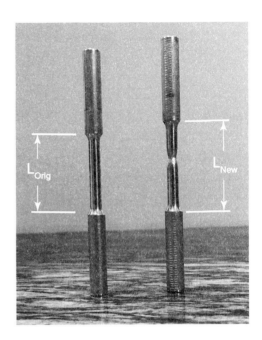

- Normal strain is expressed with the Greek letter epsilon (ε).

For most materials:

- A change in length due to a tensile force, yields a positive strain (elongation).
- A change in length due to a compressive force, yields a negative strain (shrinkage).

The equation for normal strain is:

$$\epsilon = \frac{L_{New} - L_{Original}}{L_{Original}} \text{ or } \epsilon = \frac{\delta}{L_{Original}}$$

Units for Normal Strain. Looking at the equation above, you can see that we have length divided by length. This essentially makes strain a unitless quantity. Another way to think of it is as an amount of stretch per unit of original length of the specimen.

In the example above, for every centimeter of original rubber band, there was a stretch of 1 cm, resulting in a centimeter per centimeter strain.

Poisson's Ratio

For more information regarding the Poisson's Ratio, see lesson 14.

Small Angle Assumption

When you have a bar, limb, or beam that has a small deflection (less than a few degrees), you can assume the deflected end of the beam has moved vertically down instead of through an arc. Here is a quick example:

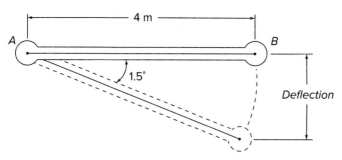

Point B at an arc

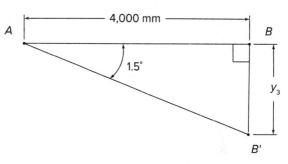

Point B assumed straight down

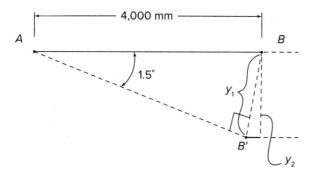

$$Sin(1.5°) = \frac{y_1}{4{,}000 \text{ mm}} \qquad y_1 = 104.71 \text{ mm}$$

$$Cos(1.5°) = \frac{y_2}{104.7078 \text{ mm}} \qquad y_2 = 104.67 \text{ mm}$$

$$Tan(1.5°) = \frac{y_3}{4{,}000 \text{ mm}} \qquad y_3 = 104.74 \text{ mm}$$

PRO TIP

Notice the very small difference between the actual value y_2 and the assumed value y_3. This assumption saves lots of time and mental struggle from figuring deflections. Thus, you can just assume that it doesn't sweep in an arc.

TEST YOURSELF 2.1

SOLUTION TO TEST YOURSELF: Strain

A heavy block (*W*) is supported by the three rods as shown. If the block moves down only vertically by 3.75 mm, what is the normal strain in the rods?

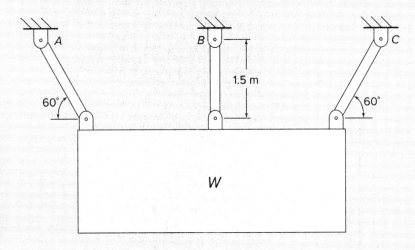

MECHANICS OF MATERIALS LESSON 8
Shear Strain Explained, Sign Convention

What Is Shear Strain?

Shear strain is given by the Greek letter gamma (γ) affectionately known by my students as "dead fish." Ha!

Shear strain is a change in angle of a body under deformation from shearing forces.

The "spring toy" is a good example of this, as you can see the individual layers of the "spring toy" sliding one relative to the next.

The "general" equation for shear strain is as follows:

$$\gamma = \frac{\pi}{2} - \theta \quad \text{(Remember } \frac{\pi}{2} \text{ is 90° in radians.)}$$

The units for shear strain are always in radians (unitless).

The reason we say this is the "general" equation is that it really only works for bodies that originally had right angle corners.

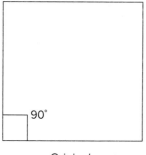

Original

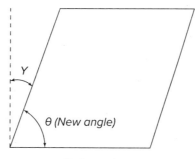

Deformed

It is better to think of shear strain as a simple change of angle. These problems can always be solved by working out angles by constructing deformation triangles.

 **PRO TIP**

Set your calculator mode to radians to solve these shear strain problems (always in radians). Don't forget to put your calculator mode back into degree mode for the rest of the problems on your homework or exam!

STRAIN

Sign Convention for Shear Strain Problems

The general convention for determining whether γ is positive or negative for a particular deformation is as follows:

- If the original angle gets smaller (example: original angle = 90°, new angle = less than 90°), the sign for γ will be positive.
- If the original angle gets larger (example: original angle = 90°, new angle = greater than 90°), the sign for γ will be negative.
- *Note:* These sign conventions can be confusing as you could have one side of a corner with a positive γ and the other side of the corner having a negative γ.

Here is a method to help you keep track of signs for γ:

- Place this on the corner of interest (i.e., at the vertex).
- If one of the sides of the angle goes into the quadrant, you look to see if that quadrant is called positive or negative. Then, you take each of these shear strain angles and add them all together (make sure you keep the sign).

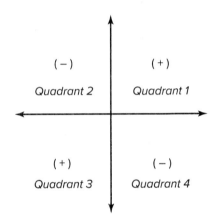

Shear Strain Sign Conventions

PITFALL

Warning: This sign convention only works for rectangular corners. It won't work for a triangular intersection. If the triangular corner gets smaller, the shear stress is positive. If it gets larger, the shear stress is negative.

STRAIN

EXAMPLE: SHEAR STRAIN

Find the shear strain at corner A (γ_A).

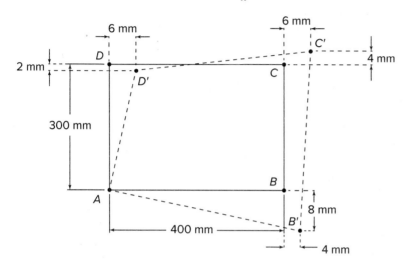

STEP 1: Determine the point of interest. This will be the corner of the specimen that we are asked for find γ for.

STEP 2: Draw the quadrant convention over the corner of interest.

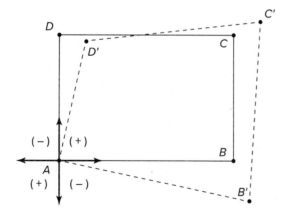

STEP 3: Determine how many of the original components of the corner have shifted. This will determine how many components are going to make up the shear strain angle γ_A. For this example, there are 2 components to γ_A which we will call γ_{A_1} and γ_{A_2}.

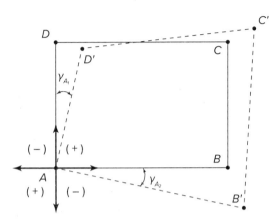

STEP 4: Determine the signs of each of the components using the quadrant convention for each of the components. For this example:

γ_{A_1} has deformed from vertical into quadrant 1, which is positive.
$\therefore \gamma_{A_1} = (+)$

γ_{A_2} has deformed into quadrant 4, which is negative.
$\therefore \gamma_{A_2} = (-)$

STEP 5: Construct triangles so that we can calculate the angle. For this example problem, we should construct two triangles—one for γ_{A_1} and one for γ_{A_2}.

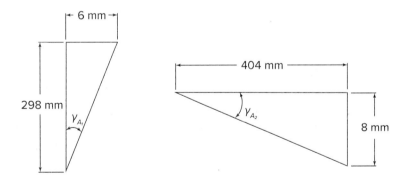

STEP 6: Calculate each component.

$$tan\ \gamma_{A_1} = \frac{6}{298} \qquad \gamma_{A_1} = 0.02013 \text{ radians}$$

$$tan\ \gamma_{A_2} = \frac{8}{404} \qquad \gamma_{A_2} = 0.01980 \text{ radians}$$

STEP 7: Apply the proper signs (which we determined earlier) to each of the components and calculate the final shear strain angle.

$$\gamma_A = \gamma_{A_1} - \gamma_{A_2} = 0.02013 - 0.01980 = 0.00033 \text{ radians}$$

Press pause on video lesson 8 once you get to the workout problem. Only press play if you get stuck.

Find the shear strain at each corner.

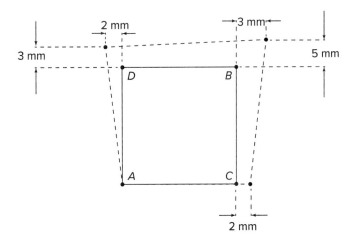

SOLUTION TO TEST YOURSELF: Strain

The rectangle shown is in uniform strain and deforms as shown. What is the strain at point *A*?

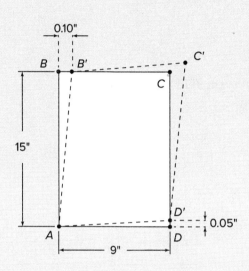

ANSWER
0.0122 radians

Strain

- Deformation that results from one of two loads: normal and shear
- Unitless property

Normal Strain (ε)

- Results from an axial load with the deformation in the direction of the applied load
- Compressive strain is negative and tensile strain is positive

Shear Strain (γ)

- Results from forces in the transverse direction with two forces of equal magnitude and opposite direction
- For rectangular corners follow the shear strain convention
- For nonorthogonal corners (i.e., triangles), if the deformation causes a smaller angle than the original angle (i.e., reduces in size) then the shear strain is positive, and if the deformation causes a larger angle then the shear strain is negative

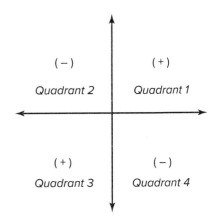

Shear Strain Sign Conventions

Equations Learned in this Level

Normal Strain

$$\epsilon = \frac{L_{New} - L_{Original}}{L_{Original}} \text{ or } \epsilon = \frac{\delta}{L_{Original}}$$

Shear Strain

$$\gamma = \frac{\pi}{2} - \theta$$

PRO TIPS

Normal Strain

- Notice the very small difference between the actual value and the assumed value for small angle deflections when swept through an arc. To save mental struggles, simply assume the deflection is straight rather than at an arc.

Shear Strain

- Set your calculator mode to radians to solve these shear strain problems (always in radians). Don't forget to put your calculator mode back into degree mode for the rest of the problems on your homework or exam!

PITFALL

Shear Strain

Warning: This sign convention only works for rectangular corners. It won't work for a triangular intersection. If the triangular corner gets smaller, the shear stress is positive. If it gets larger, the shear stress is negative.

Level 3

Mechanical Properties of Materials

 **MECHANICS OF MATERIALS LESSON 9**
Stress-Strain Diagram, Guaranteed for Exam 1!

Introduction

The entire goal of mechanics of materials is to find internal stresses for loads applied to predict failure (or design so that it doesn't fail). Failure can entail exceeding the strength of a particular material, or excessive deformation. Thus, it is important to understand material properties. All materials, including alloys of the same material, exhibit different properties.

PRO TIP | Hopefully, you are using this book as an additional resource for your Mechanics of Materials class. This is the *one section* of your course that you need to read the entire chapter. There are a lot of terms and information in this level. It is highly recommended to make flash cards (yes this is old-school), as it is definitely a technique that has been proven to work!

The Stress-Strain Diagram

Where do all the values for these material properties come from?

Answer: They are "empirically" derived or observed. This means that each material is tested and the results are plotted.

The Tensile Test

Most material properties are observed by placing a specimen in a tensile tester and applying a load to the specimen until failure.

Process:

- As the lead screw is rotated, a load is applied to the "dogbone" specimen and it begins to elongate. (**Note:** Some machines are hydraulic and do not use a screw.)
- As elongation continues, the specimen begins to "neck" or reduce in cross section area until it finally fractures.

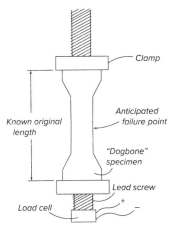

Clamp

Known original length

Anticipated failure point

"Dogbone" specimen

Lead screw

Load cell

- The output is force via a load cell and elongation (typically via an extensometer).
- The load is converted to stress and the elongation is converted to strain.

ACTUAL STRESS-STRAIN DIAGRAM

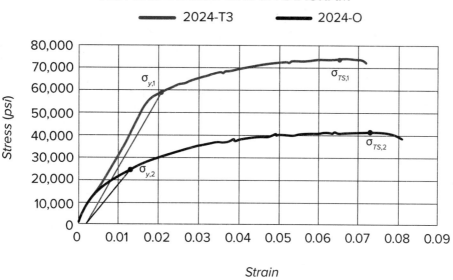

This is an example of the stress versus strain diagram generated from a tensile test.

STRESS-STRAIN DIAGRAM FOR DUCTILE MATERIAL

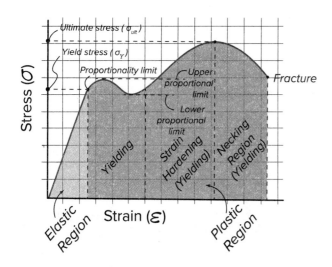

Elastic   Yielding   Necking   Fracture

Stress: Intensity of a force ($\sigma = \dfrac{P}{A}$)

Strain: Deformation or Elongation typically normalized to an original dimension ($\varepsilon = \dfrac{\delta_L}{L_0}$)

Why do we plot stress versus strain instead of force versus elongation?

We plot stress versus strain instead of load versus elongation as we don't want to have to look up the force that the sample can withstand for every single change in geometry. What a pain! Thus, we plot the stress intensity so it is not dependent on the geometry of the cross-section.

Elastic Region

At any point along this part of the curve, once the load is removed from the specimen, it will return to its original length. The curve in this region is completely linear.

Elastic Modulus (Modulus of Elasticity) Material

Describes the elasticity or the "bendiness" of a material.

- The ability to deform and then return to the original position (bonds are stretching).
- $E = \sigma/\varepsilon$ (slope of the linear portion of the stress-strain curve, i.e., rise/run).
- *Only* valid for stress values below yield stress (σ_Y).
- Units are the same as stress.

Increasing Stiffness

Stress (σ) vs Strain (ε)

Note: Robert Hooke figured out that force and deformation are proportional to each other (sometimes the proportionality constant is given as *k*). This was a huge breakthrough at the time! Then, later on, we developed a more useful way to describe materials in terms of force divided by cross-sectional area (i.e., stress) versus strain. This proportionality constant is the elastic modulus.

$F = kx$ $\sigma = E\varepsilon$

(Hooke's law)   (where *E* is the Elastic Modulus or sometimes called the Young's Modulus)

CHALLENGE QUESTION

Most tensile testing of metals is conducted with 0.505 inch diameter bars. If you test 0.400 inch diameter steel bars, the elastic modulus is likely to be (A. higher, B. lower, or C. the same as) the modulus found using the 0.505 inch diameter steel bars.

CHALLENGE QUESTION ANSWER
Same as (material properties are not geometry dependent)

MECHANICAL PROPERTIES OF MATERIALS

Plastic Region

"Plastic" in this instance means that once the load is removed from the specimen, there will be permanent deformation.

- **Yield Stress.** The point on the diagram that transitions from elastic to plastic.
  - Any loads above yield stress (σ_Y) will result in permanent deformation.
- **Proportional Limit.** The highest point on the diagram where stress and strain are directly proportional (linear portion).
  - Upper and Lower Proportional Limit. Range where slipping and molecular shifts within the material take place, causing unique deformation without a change in stress.
  - *Note:* Yield stress and the proportional limit occur so closely together that they are usually considered to be the same. Thus, from here forward we will only be using yield stress (σ_Y) for this point on the diagram.
- **Strain Hardening.** Strengthening of material due to plastic deformation (you might have heard of cold-working, which is the exact same thing).

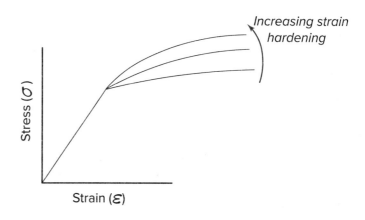

- **Necking.** As the load increases, the cross section begins to reduce.
- **Ultimate Stress.** The maximum stress where a material transitions between strain hardening and necking.
- **Fracture.** Material specimen breaks and separates into two pieces.

PRO TIP

When designing with ductile steels, sometimes an assumption is that the material behaves in an elastic-perfectly plastic manner (ignoring strain hardening). This would *idealize* the stress versus strain curve as a flatline beyond the yield point.

Other material properties determined from the stress versus strain diagram:

- **Resilience.** The ability of a material to store energy when deformed elastically (the "modulus of bounce-backiness").
- **Toughness.** The ability of a material to absorb energy until failure.

Note: Both of these will be discussed in Lesson 13.

True Stress-Strain Diagram Versus Engineering

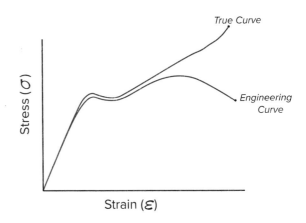

- **Engineering curve.** This curve is the one most commonly used. This curve does *not* consider the change in cross section as necking occurs.
- **True curve.** This is what actually happens if you could track the change in cross section as necking occurs.

Ductile Versus Brittle

In this course, we typically will classify materials into two categories:

- **Ductile.** The ability to be deformed permanently without fracturing (planar slip). Think of a metal pipe—typically it bulges before failure so tons of warning.
  - **Examples:** Aluminum, copper, steel, brass, bronze, titanium, zinc (i.e., most metals), mozzarella cheese
- **Brittle.** Materials where there is a very small amount of deformation before fracturing (think of a clay pipe breaking—there is very little warning before shattering)
  - **Examples:** Glass, ceramics, cast iron, acrylic, timber, peanut candy

Example of cup and cone failure typical of ductile material failures.

Materials are classified as ductile or brittle based on their percent elongation before fracture (i.e., percent plastic deformation). Typically ductile materials have percent EL greater than 5%.

Note: Brittle materials, such as concrete or cast iron, behave differently in tension than they do in compression. Generally, they are much stronger in compression than they are in tension. This is why we often "pre-tension" the steel reinforcement in concrete to place the concrete into compression (as is done in bridges) as concrete is great in compression but weak in tension.

TEST YOURSELF 3.1

SOLUTION TO TEST YOURSELF: Mechanical Properties of Materials

1. Two bars both alike in geometry, in Texas where we set our scene (Shakespeare—get it—lol). Bar *A* deflects more than Bar *B*. When the load is removed, both bars return to the original length. From this we can conclude that Bar *A* is _____. (Circle all that apply.)
   a. stronger   b. weaker   c. stiffer   d. less stiff   e. more ductile
   f. less ductile   g. can't conclude from given

2. Bar *A* and Bar *B* are loaded identically until one bar fails. The bar that fails first is _____.
   (Circle all that apply.)
   a. stronger   b. weaker   c. stiffer   d. less stiff   e. more ductile
   f. less ductile   g. can't conclude from given information

3. For a bulletproof vest, which material property would you be most interested in?
   a. yield strength   b. ultimate tensile strength   c. modulus of resilience
   d. modulus of toughness

ANSWERS
1. (d) less stiff
2. (b) weaker
3. (d) modulus of toughness

MECHANICS OF MATERIALS LESSON 10
0.2% Offset Rule Explained, Yield Point

What Is the 0.2% Offset Rule?

The 0.2% offset rule is used to estimate yield stress in brittle materials as well as some ductile materials like aluminum because it is difficult to tell exactly where this curve switches from linear to nonlinear.

Most ductile materials have stress-strain curves where it's easy to identify where the curve goes from being linear to nonlinear; this may not be the case for brittle materials.

SAMPLE OF BRITTLE MATERIALS

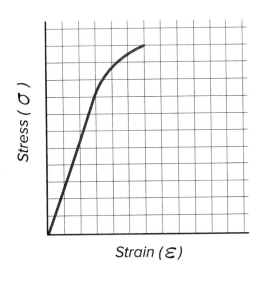

SAMPLE OF DUCTILE MATERIALS

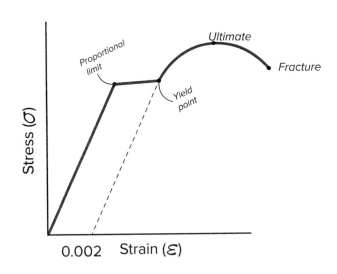

PRO TIP

The stress-strain diagrams you see in textbooks often use different scales on the horizontal axis for different parts of the graph. Always pay attention to changes in units on these axes.

MECHANICAL PROPERTIES OF MATERIALS

How to Use the 0.2% Offset Rule Recipe

STEP 1: Convert 0.2% to decimal (0.002) and identify that point on the (ε) strain axis.

STEP 2: Construct a parallel line to the linear portion of the stress-strain curve. (It's easiest to count squares in two different places and then connect the points to construct the parallel lines.)

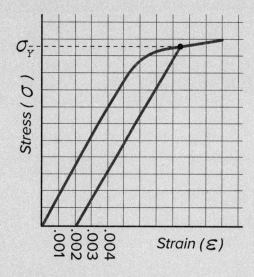

STEP 3: Estimate yield stress by reading off the value of stress at that intersection point.

Note: Since this is just an estimation of yield stress, some industries, such as aerospace, may choose to use an estimation that allows for tighter tolerances. Those industries may use a 0.1% offset or a 0.15% offset rule.

PITFALL

Don't forget it is 0.2% so 0.002! I have seen several students call this the 2% instead of 0.2% offset rule and use this technique but put the intercept at 0.02 which is wrong!

WATCH VIDEO **MECHANICS OF MATERIALS LESSON 11**
Modulus of Elasticity Example Problem

▶ Press pause on video lesson 11 once you get to the workout problem. Only press play if you get stuck.

DA is rigid. If *W* causes point *B* to displace downwards 0.025", find the strain in *DE* and *BC* and find *W*. The elastic modulus of A-36 steel is 29×10^6 psi. Note in this book CSA means cross-sectional area.

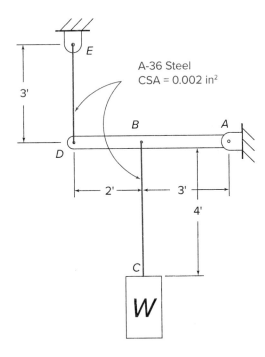

A-36 Steel
CSA = 0.002 in²

TEST YOURSELF 3.2

SOLUTION TO TEST YOURSELF: Mechanical Properties of Materials

For the stress versus strain curve below, compute the Elastic Modulus and the yield stress using the 0.2% offset rule.

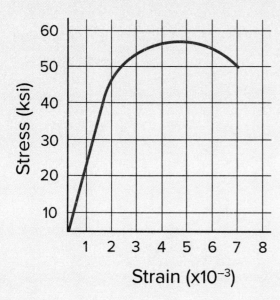

ANSWERS
$E \approx 22 \times 10^6$ psi
Yield stress ≈ 56 ksi

✓ TEST YOURSELF 3.3

SOLUTION TO TEST YOURSELF: Mechanical Properties of Materials

Bar *AB* is rigid and supported by a pin connection at the left end and by two metal rods (*CD* and *EF*) each with a cross-sectional area of 1.5 in². If point *B* displaces downward 0.070 in, compute the normal strain in the two cables and the weight *W*. The cables are made from A-36 steel. The elastic modulus of A-36 steel is 29×10^6 psi.

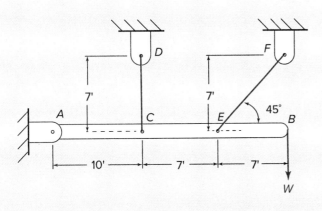

ANSWERS

$W = 12{,}730$ lb

$\varepsilon_{CD} = 3.473 \ 10^{-4}$ in/in

$\varepsilon_{EF} = 2.951 \ 10^{-4}$ in/in

MECHANICS OF MATERIALS LESSON 12
Poisson's Ratio Example Problem, Using Strain

Poisson's Ratio

- Poisson's Ratio (poisson in French means "fish") is the ratio of an object's change in length to its change in diameter.
- Think about an object being composed of a certain number of molecules, and when an object stretches under load, the material to increase the length has to come from somewhere—thus, the reduction in the diameter.
- I always think of a marshmallow! If you "squish" the length, you get a change in the diameter.

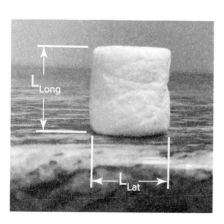 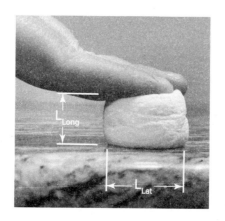

Equation for Calculating Poisson's Ratio

- Given the Greek letter for nu (ν)
- $$\nu = -\frac{\varepsilon_{long}}{\varepsilon_{lat}}$$
  - ε_{long}. Think of this as along the axis of the specimen (axial dimension). Also in the direction of the force applied.
  - ε_{lat}. This is perpendicular to the force applied.
- If you have a round specimen, you only get a change in diameter.
- If you have a rectangular specimen, you get a change in both x and y.

But why is there a negative in front of Poisson's equation?

- If there is a positive growth in one direction, there is nearly always a negative growth in the perpendicular direction. A negative divided by a positive will give you a negative. That negative sign in front of the equation will then turn Poisson's Ratio back into a positive sign.
- Poisson's Ratio is nearly always positive. It ranges between 0 and about 0.5 and is a *unitless* value.

  Example:

  Cork = 0

  A-36 Steel = 0.32

  Aluminum 6061-T6 = 0.33

  Rubber = 0.5

  (Only in really weird materials, such as auxetic materials, are Poisson's Ratio negative.)

PRO TIP

How do you know which stress is longitudinal versus lateral? The longitudinal is always in the direction of the applied force, and the lateral is always the one perpendicular to that.

If you are a football fan, then an easy way to remember this is that when you throw a lateral you are throwing to the side.

EXAMPLE: POISSON'S RATIO

A rectangular bar has dimensions as shown. Determine its change in length and its new cross-sectional dimensions while under load. The material behaves elastically. $E_{ST} = 200$ GPa and $\nu = 0.32$.

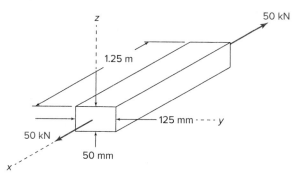

STEP 1: Find the normal stress in the bar.

$$\sigma_y = \frac{P}{A} = \frac{50{,}000 \text{ N}}{(50)(125)\text{mm}^2} = 8 \; MPa$$

STEP 2: Calculate the normal strain in the bar (axial).

$$\varepsilon_y = \frac{\sigma_y}{E_{ST}} = \frac{8 \dfrac{\text{N}}{\text{mm}^2}}{2{,}000{,}000 \dfrac{\text{N}}{\text{mm}^2}} = 0.00004 \; \frac{\text{mm}}{\text{mm}}$$

STEP 3: Use Poisson's Ratio to calculate the new cross-section dimensions.
Note: The change in the y-direction is the same as the change in the z-direction.

$$\nu = -\frac{\varepsilon_{lat}}{\varepsilon_{long}} \Rightarrow \nu(\varepsilon_{long}) = -\varepsilon_{lat}$$

$$0.32(0.00004) = -\varepsilon_{lat}$$
$$1.28 \times 10^{-3} \text{ mm} = -\varepsilon_{lat}$$

STEP 4: Calculate new dimensions by subtracting the change (δ) from the original length.

$$\delta_z = \varepsilon(L_z) = -(0.00128 \; \frac{\text{mm}}{\text{mm}})(50 \text{ mm}) = -0.064 \text{ mm}$$

$$\delta_y = \varepsilon(L_y) = -(0.00128 \; \frac{\text{mm}}{\text{mm}})(125 \text{ mm}) = -0.16 \text{ mm}$$

New cross-section dimensions:

z dim: 50 mm − 0.064 mm = 49.936 mm

y dim: 125 mm − 0.16 mm = 124.84 mm

MECHANICAL PROPERTIES OF MATERIALS

▶ Press pause on video lesson 12 once you get to the workout problem. Only press play if you get stuck.

If the beam is to remain horizontal, find the distance x where the 80 kN force must be located and the new diameter of column A.

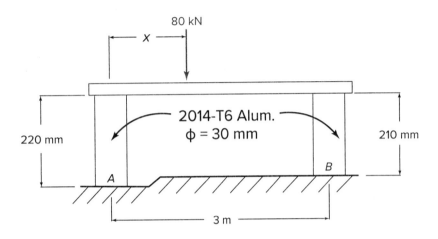

TEST YOURSELF 3.4

SOLUTION TO TEST YOURSELF: Mechanical Properties of Materials

1. For the frame shown in Test Yourself 3.3, what is the change in diameter of rod *CD* if the Poisson's Ratio is 0.32.

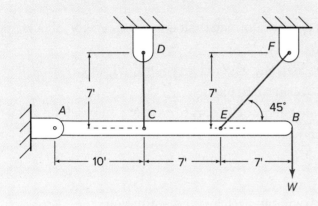

2. A flat plate is uniformly loaded as shown. A circle of radius *r* is lightly etched in the center of the plate. When loaded, it is observed that the circle deformed into an ellipse with dimensions *a* and *b* as indicated.

(A) Develop an equation for finding the Poisson's Ratio from this information. (**Hint:** Use your basic definitions of normal strain in the *x* and *y* directions along with the definition of Poisson's Ratio.)

(B) If $r = 1.00"$, $b = 1.94"$, and $a = 2.20"$, what is Poisson's Ratio?

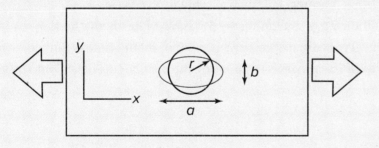

ANSWERS

1. Diameter change = −0.000154 in

2A. $\nu = -(b - 2r)/(a - 2r)$

2B. $\nu = 0.30$

MECHANICS OF MATERIALS LESSON 13
Strain Energy; Example Problems from Stress-Strain Diagram

Strain Energy

Strain energy is the energy stored in a material due to deformation. It is a broad category that includes both modulus of resilience and modulus of toughness.

Think of any material as a spring. When you load any material, as you would load a spring, that deformation stores that energy in the material, called strain energy.

Design engineers take advantage of the strain energy behaviors of different materials for things such as car bumpers, armor, or bulletproof vests.

Modulus of Resilience

- Amount of energy a material can absorb without permanent deformation.
- Calculated by taking the area under the elastic portion of the stress-strain diagram. Since the area under the elastic portion of the stress-strain curve is a triangle, the equation is as follows:
  $M_R = (\frac{1}{2})\sigma_y \varepsilon$
- Units are ksi or MPa.

Modulus of Toughness

- Total amount of energy a specimen can absorb before fracture.
- Very important for materials that "cannot" fail, such as a bulletproof vest.
- Total area under the stress-strain curve, including the Modulus of Resilience area.
  Note: Since the stress-strain curve is not defined by a particular mathematical function, we have to estimate the Modulus of Toughness by estimating the number of "boxes" under the curve (one "box" is equal to $\sigma \times \varepsilon$).
- Units are ksi or MPa.

Calculating Strain Energy

If you aren't sure how to convert from stress to energy, just multiply by the volume of the specimen (unit analysis—gotta love it!).

$$\frac{N}{mm^2}(mm^3) = N(mm) \quad OR \quad \frac{kips}{in^2}(in^3) = kip(in)$$

- **Modulus of Resilience.** The amount of energy a specimen can absorb without permanent deformation.
  $E_R = M_R(V)$ M_R = Modulus of Resilience   V = Volume of Specimen
- **Modulus of Toughness.** The amount of energy a specimen can absorb before fracture.
  $E_T = M_T(V)$ M_T = Modulus of Toughness   V = Volume of Specimen

EXAMPLE: STRESS VERSUS STRAIN DIAGRAM

For the given stress versus strain curve, find the following:

A. Modulus of Elasticity
B. Modulus of Resilience
C. Yield Stress
D. Ultimate Tensile Stress
E. Modulus of Toughness
F. The total amount of energy the given specimen can absorb without permanent deformation
G. The total amount of energy the specimen can absorb before fracture

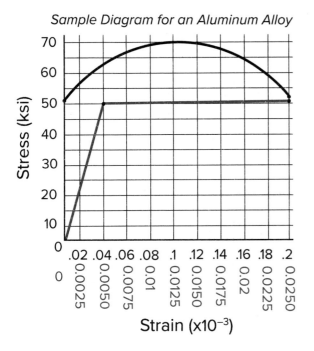

Sample Diagram for an Aluminum Alloy

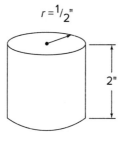

A. Modulus of Elasticity is the slope of the linear portion of the stress versus strain curve.

$$E = \frac{\sigma}{\varepsilon} = \frac{50}{0.005} = 10{,}000 \text{ ksi or } 10 \times 10^3 \text{ ksi}$$

B. Modulus of Resilience is the area under the linear position of the stress versus strain curve.

$$M_R = (\tfrac{1}{2})\sigma_y\varepsilon = (\tfrac{1}{2})(50)(0.005) = 0.125 \text{ ksi}$$

C. Yield stress is the onset of permanent deformation. Sometimes it is approximated as the location where the curve changes from linear to nonlinear (from observation).
Note: Sometimes a 0.2% offset method is used. For this example, $\sigma_y = 50$ ksi.

D. Ultimate stress is the highest stress on the stress vs strain diagram: $\sigma_{UTS} = 70$ ksi.

E. Toughness is the total area under the curve. *Note:* It can be approximated by counting boxes.

bottom curve (zoomed-in one): 50.5 boxes $\times$ (10 ksi $\times$ 0.0025) = 1.2675 ksi

top curve (zoomed-out one): 70 boxes $\times$ (10 ksi $\times$ 0.02) = 14 ksi

$$M_T = \text{total} = 1.2675 + 14 = 15.26 \text{ ksi}$$

F. $E_R = M_R(V) = 0.125$ ksi $\times \left(\pi(0.5 \text{ in})^2 \times 2.5 \text{ in}\right) = 0.245$ kip-in

G. $E_T = M_T(V) = 15.26$ ksi $\times \left(\pi(0.5 \text{ in})^2 \times 2.5 \text{ in}\right) = 29.96$ kip-in

Press pause on video lesson 13 once you get to the workout problem. Only press play if you get stuck.

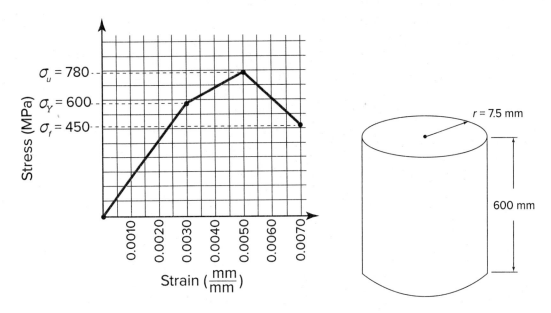

Find the following quantities for the specimen below with the stress-strain diagram shown. Note that the Poisson's Ratio is 0.35.

A. What is the elastic modulus?

B. Find the largest force you can apply without deformation if a factor of safety of 2 is desired.

C. What max tensile load can the rod sustain before fracture?

D. While under a tensile load of 2.25×10^4 N, the rod will elongate how much?

E. While under a tensile load of 2.25×10^4 N, the rod's diameter will decrease to what?

F. What is the maximum energy sustained before fracturing?

G. What is the maximum energy sustained without permanent deformation?

H. If loaded to 12.4×10^4 N and then unloaded, find the permanent deformation in length.

> ▶ WATCH VIDEO   **MECHANICS OF MATERIALS LESSON 14**
> **Stress-Strain Diagram Example Problem**

▶ Press pause on video lesson 14 once you get to the workout problem. Only press play if you get stuck.

Find the following quantities for the specimen below with the stress-strain diagram shown.

   A. Modulus of Resilience
   B. Modulus of Elasticity
   C. Modulus of Toughness
   D. The new length of the specimen if 1,000# is applied.

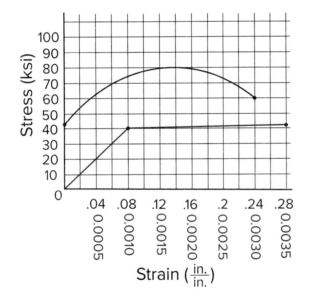

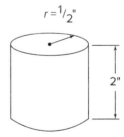

$r = {}^{1}/_{2}"$

2"

TEST YOURSELF 3.5

SOLUTION TO TEST YOURSELF: Mechanical Properties of Materials

Find the Modulus of Resilience, Elastic Modulus, and the Modulus of Toughness from the stress-strain diagram shown below. Note that the material yields at 40 ksi.

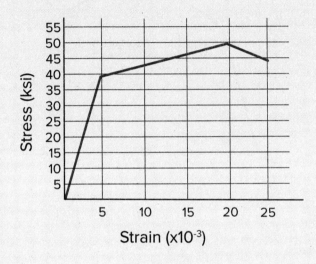

ANSWERS

Elastic Modulus = 8×10^6 psi

Modulus of Resilience ≈ 100 psi

Modulus of Toughness ≈ 1,010 psi

MECHANICS OF MATERIALS LESSON 15
Shear-Stress Strain Diagram

Shear Stress-Strain Diagram

- Axes are shear stress (τ) and shear train (γ).
  - Recall: $\tau = \dfrac{V}{A}$ and $\gamma = \dfrac{\pi}{2} - \theta$

- Linear portion (exactly like the stress-strain) leads to the Shear Modulus of Elasticity (G), also called the Shear Modulus of Rigidity.

- Shear Modulus of Elasticity: $G = \dfrac{\tau}{\gamma}$

  - Units for the Shear Modulus of Elasticity (G) are units of shear stress (MPa or ksi) since the units for gamma are radians which are unitless.

  - *Note:* In addition to the equation listed above, there is another way to solve for G using the material constants of E and ν: $G = \dfrac{E}{2(1 + \nu)}$.

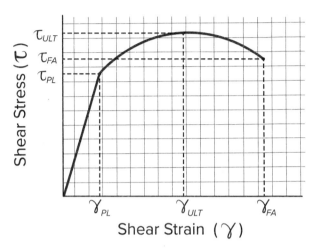

PRO TIP

If you know the value of two of these variables (*G*, *E*, or ν), you can find the third via the relationship above. We can't independently assign all three randomly as then it would invalidate the equation above.

MECHANICAL PROPERTIES OF MATERIALS

MECHANICS OF MATERIALS LESSON 16
Creep and Fatigue with the S-N Diagram

Other Material Failure Modes

Creep and fatigue are covered here for the sole purpose of conceptually understanding their existence. Typically, these kinds of questions will only show up as conceptual questions.

Creep

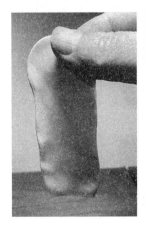
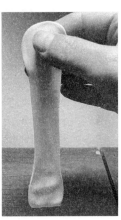

- Failure mode for materials that are continuously loaded and typically at elevated temperatures.
- A material, when loaded, will deform over extremely long periods of time.
- Example: boiler components might fail from creep around 100,000 hours.
- An easy way to wrap your mind around the concept of creep is this example of children's putty. When it's shaped into a thin slab (left), the putty will slowly deform under its own weight (right). This is a greatly sped-up example of watching creep in action.

Fatigue

- Type of failure that comes from cyclical loading and unloading.
- Ability of a material to be flexed or plastically deformed over and over without breaking.
- Example: Plastic hinge seen on shampoo bottles and toothpaste tubes—called a living hinge. It is specifically designed to bend back and forth "essentially" an infinite number of times. So far, no one has been able to test an infinite number of cycles.

The S-N Diagram

- Compares stress amplitude or load applied to a material specimen cyclically over and over again.
- There is a stress limit for certain materials, where no matter how many cycles the material undergoes, the material will not fail in fatigue.
- Example: as long as the steel specimen in the graph is never cyclically loaded more than 28 ksi, it can be loaded and unloaded an infinite number of times without failure.

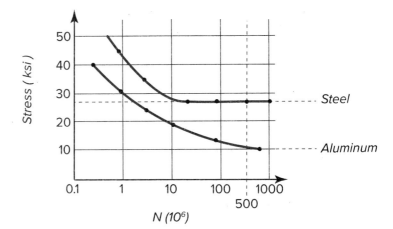

Material Properties

- Most often in mechanics of materials we consider something as "failed" when it doesn't return back to its original shape when unloaded. The property that defines this limit is called the yield stress. Sometimes this is found via a 0.02% offset method.
- Another material property used often is the Young's Modulus (sometimes called the elastic modulus). This is the slope on the stress-strain curve prior to yielding. It describes the stiffness of a material.
- There are other mechanical properties of materials on a stress-strain diagram that are sometimes used in specific applications such as Modulus of Resilience, Modulus of Toughness, and the ultimate tensile stress.

Stress Strain Diagrams

- These diagrams relate the change in length (deformation) as the load (normal stress) is increased. These diagrams are observed, not mathematically derived and tell us properties of the material such as:
  - Modulus of Elasticity
  - Yield Stress
  - Ultimate Stress
  - Fracture Stress
- We also looked at the differences in ductile and brittle material. Both materials have a linear (elastic) portion of the curve, but ductile materials have much more deformation before fracture occurs.

Strain Energy

- Strain energy is the energy stored in a material due to deformation. It is a broad category that includes both the Modulus of Resilience and Modulus of Toughness. We use these "moduli" to determine how much energy materials can absorb without permanent deformation (Modulus of Resilience) or without fracture (Modulus of Toughness).

Shear Stress—Shear Strain Diagrams

- These diagrams relate the change in shear strain (change in angle in radians) as shear load (shear stress) is increased. These diagrams are observed, not mathematically derived and tell us properties of the material such as the Shear Modulus of Resilience.

Poisson's Ratio

- Describes the dimensional changes of an object under an applied load.
- Poisson's Ratio is nearly always positive. It typically ranges between 0 and about 0.5 and is a unitless value.
- When an object elongates, there is a change in the cross-section.
  - If you have a round specimen, you only get a change in diameter.
  - If you have a rectangular specimen, you get a change in both x and y.
  - This was only an introduction to the concept of Poisson's Ratio. Workout problems will be in future levels.

Failure Modes

- Creep is the change in length of a material without increasing the load and typically occurs at elevated temperatures.
- Fatigue is the cyclical loading and unloading of a material.
- S-N curve is a plot of the number of cycles until failure when at a specified stress amplitude (the mean stress is the same for all samples tested).

Equations Learned in this Level

Young's Modulus

$$E = \frac{\sigma}{\varepsilon}$$

Poisson's Ratio

$$\nu = -\frac{\varepsilon_{long}}{\varepsilon_{lat}}$$

Shear Modulus of Elasticity

$$G = \frac{\tau}{\gamma}$$

$$G = \frac{E}{2(1 + \nu)}$$

Shear Strain

$$\gamma = \frac{\pi}{2} - \theta$$

Modulus of Resilience

$$M_R = (\tfrac{1}{2})\sigma_y\varepsilon$$

PRO TIPS

Stress-Strain Diagram

- Hopefully, you are using this book as an additional resource for your Mechanics of Materials class. This is the *one section* of your course that you need to read the entire chapter. There are a lot of terms and information in this level. It is highly recommended to make flash cards (yes this is old-school), as it is definitely a technique that has been proven to work!
- When designing with ductile steels, sometimes an assumption is that the material behaves in an elastic-perfectly plastic manner (ignoring strain hardening). This would *idealize* the stress versus strain curve as a flatline beyond the yield point.
- The stress-strain diagrams you see in textbooks often use different scales on the horizontal axis for different parts of the graph. Always pay attention to changes in units on these axes.

Poisson's Ratio

- How do you know which stress is longitudinal versus lateral? The longitudinal is always in the direction of the applied force, and the lateral is always the one perpendicular to that.

     If you are a football fan, then an easy way to remember this is that when you throw a lateral you are throwing to the side.

Shear Stress-Strain Diagram

- If you know the value of two of these variables (*G*, *E*, or ν), you can find the third via the relationship above. We can't independently assign all three randomly as then it would invalidate the equation above.

PITFALL

Stress-Strain Diagram
- Don't forget it is 0.2% so 0.002! I have seen several students call this the 2% instead of 0.2% offset rule and use this technique but put the intercept at 0.02 which is wrong!

Level 4
Axial Stress and PL/AE

 WATCH VIDEO **MECHANICS OF MATERIALS LESSON 17**
Axial Elongation Due to Axial Load Example

Axial Elongation

- Specimens elongate or deform due to axial loads or due to change in temperature.
  - Changes due to tensile loads are positive (+) (remember tension starts with a plus)
  - Changes from compression loads are negative (−)
- To compute the amount of axial elongation or reduction (δ):

  $\delta = \dfrac{PL}{AE}$ = Change in length (can be positive or negative)

  P = the axial force applied along the length (normal force [N])

  L = the *original* length of the specimen

  A = the cross-sectional area of the specimen

  E = Modulus of Elasticity

Note: My students commonly refer to this as the "play" equation. . . . Maybe that will help you remember the equation!

PRO TIP

This is a bit of a repeat, and probably not the last. When using the "play" equation, always include your units in your work as each term in the equation will present you with an opportunity for mixed units! Break your units down to basic units like **N** and **mm** for metric or **kips** and **inches** for Imperial units.

PITFALL

A common mistake made when using the "play" equation is which length to use for the L in the equation. It is never the stretched or elongated length, but *always* the original length (L_0) of the specimen!

Make sure to plug in the proper "sign" of the force: positive if in tension and negative if in compression. This is extremely critical if your shaft has multiple forces being applied at each section.

EXAMPLE: AXIAL ELONGATION

For the following stepped shaft, calculate the total displacement of end A with respect to end D. $E_{ST} = 200$ GPa, $E_{AL} = 70$ GPa, $E_{CU} = 110$ GPa, $\Phi_{AB} = 50$ mm, $\Phi_{BC} = 100$ mm, $\Phi_{CD} = 30$ mm.

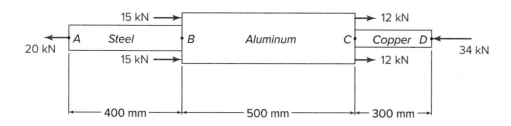

STEP 1: Calculate the force in each section of the stepped shaft. **Recall:** To calculate the internal force within a section, cut through the section of interest and draw the FBD of the easiest side to find the internal force (N). Also, determine if that section is in tension or compression.

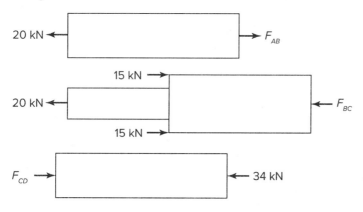

$\Sigma F_X = 0 = N_{AB} - 20$ kN $\quad\quad$ $\Sigma F_X = 0 = 15$ kN $+ 15$ kN $- 20$ kN $- N_{BC}$ $\quad\quad$ $\Sigma F_X = 0 = N_{CD} - 34$ kN

$N_{AB} = 20$ kN (T) $\quad\quad\quad\quad\quad$ $N_{BC} = 10$ kN (C) $\quad\quad\quad\quad\quad\quad\quad\quad\quad\quad\quad\quad$ $N_{CD} = 34$ kN (C)

$N_{AB} = 20{,}000$ N (T) $\quad\quad\quad\quad$ $N_{BC} = 10{,}000$ N (C) $\quad\quad\quad\quad\quad\quad\quad\quad\quad\quad\quad$ $N_{CD} = 34{,}000$ N(C)

Note: You could also analyze the left side of the beam and cut thru CD. You will get the same answer.

$\Sigma F_X = 0 = -20$ kN $+ 15$ kN $+15$ kN $+ 12$ kN $+ 12$ kN $- N_{CD}$

$N_{CD} = 34$ kN (C)

AXIAL STRESS AND PL/AE

STEP 2: Compute cross-section area for each section. ***Danger:*** Don't use diameter where you should be using radius!

$A_{AB} = \pi\,(25\text{ mm})^2 = 1{,}963.5\text{ mm}^2$

$A_{BC} = \pi\,(50\text{ mm})^2 = 7{,}853.98\text{ mm}^2$

$A_{CD} = \pi\,(15\text{ mm})^2 = 706.86\text{ mm}^2$

STEP 3: Identify remaining components of the equation, L and E, and convert them to basic units, in this case, Newtons (N) and millimeters (mm).

$L_{AB} = 400\text{ mm}$ $E_{ST} = 200\text{ GPa or }200{,}000\text{ MPa or }200{,}000\,\dfrac{\text{N}}{\text{mm}^2}$

$L_{BC} = 500\text{ mm}$ $E_{AL} = 70\text{ GPa or }70{,}000\text{ MPa or }70{,}000\,\dfrac{\text{N}}{\text{mm}^2}$

$L_{CD} = 30\,0\text{ mm}$ $E_{CU} = 110\text{ GPa or }110{,}000\text{ MPa or }110{,}000\,\dfrac{\text{N}}{\text{mm}^2}$

STEP 4: Determine the elongation using the "play" equation.

$$\delta_{AD} = \frac{(20{,}000\text{ N})(400\text{ mm})}{(1{,}963.5\text{ mm}^2)(200{,}000\,\frac{\text{N}}{\text{mm}^2})} - \frac{(10{,}000\text{ N})(500\text{ mm})}{(7{,}853.98\text{ mm}^2)(70{,}000\,\frac{\text{N}}{\text{mm}^2})} - \frac{(34{,}000\text{ N})(300\text{ mm})}{(706.86\text{ mm}^2)(110{,}000\,\frac{\text{N}}{\text{mm}^2})}$$

$\delta_{AD} = 0.0204\text{ mm} - 0.0090\text{ mm} - 0.131\text{ mm}$

$\delta_{AD} = -0.120\text{ mm}$

The negative indicates that the stepped shaft has an overall reduction in total length of 0.120 mm.

▶ Press pause on video lesson 17 once you get to the workout problem. Only press play if you get stuck.

F ind the change in length of a 50 mm diameter bar looking at *A* with respect to *E*. The bar is composed of 2014-T6 aluminum with a Young's Modulus of 73.1 GPa.

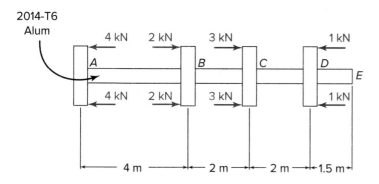

TEST YOURSELF 4.1

SOLUTION
TO TEST
YOURSELF:
Axial Load
(PL/AE)

Rod *A* has a diameter of 50 mm and length 350 mm. It has an upward force
$P = 45$ kN acting on the top, while the bottom is fastened to a rigid plate (*C*).
Two rods (*B*) on either side each have a diameter of 15 mm and length of 275 mm. What is the
displacement of the top of Rod *A* due to this loading? Use $E = 70$ GPa for all rods.

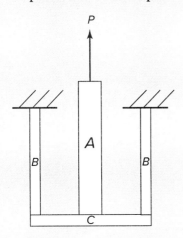

 WATCH VIDEO

MECHANICS OF MATERIALS LESSON 18
Axial Elongation Example Problem, Displacement

Press pause on video lesson 18 once you get to the workout problem. Only press play if you get stuck.

Find the displacement of points *E*, *B*, and *C* if *CED* and *AB* are rigid bars. Links *FC* and *EB* are 25 mm in diameter. The links are both composed of A-36 steel with a Young's modulus of 29,000 kip/in².

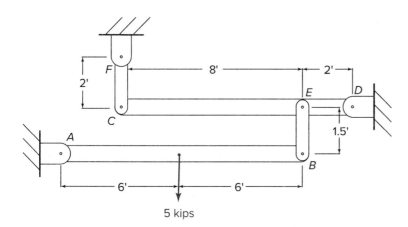

AXIAL STRESS AND PL/AE

✓ TEST YOURSELF 4.2

SOLUTION TO TEST YOURSELF: Axial Load (PL/AE)

Rigid bar *ABC* is loaded as shown. A 4-ft steel wire, *DB*, with diameter $\frac{1}{8}$ in is also attached to the bar. What is the displacement of points *B* and *C* due to this loading? Take $E = 29 \times 10^6$ psi.

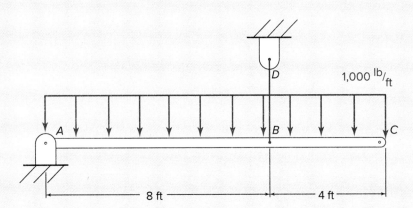

$$1,000 \, ^{lb}/_{ft}$$

8 ft   4 ft

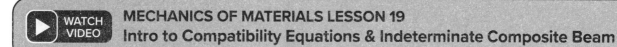

Statically Indeterminate Problems

- **Statically determinate problems** can be solved simply through equilibrium equations ($\Sigma F_x = 0$, $\Sigma F_y = 0$, $\Sigma F_z = 0$, and $\Sigma M_x = 0$, $\Sigma M_y = 0$, $\Sigma M_z = 0$).
- **Statically indeterminate problems** are those where the equations of equilibrium, by themselves, are not enough to solve for the internal forces because there are not enough equations for the number of unknowns. These systems are supplemented with additional equations called compatibility equations so a solution can be reached.
- **Compatibility equations** are the additional equations needed to solve for statically indeterminate systems. These equations come from a geometry constraint.

Statically indeterminate problem types that you will likely come across are:

- **Dependent members.** Beams with different links or two-force members supporting them. Compatibility equations are visually derived from similar triangle relationships (see Test Yourself 4.4).
- **Composite materials problems.** Systems composed of multiple materials undergoing an axial load. Compatibility equations are generally the change in length of one material is the same as the change in the other material. In addition, we also see a sharing of the load by the two materials, which usually leads to an additional compatibility equation (see Video 19: Composite Material Problem).
- **Specimen between two walls or supports and a wall.** Compatibility equations are derived from a superposition technique where we ignore the wall and then use the reaction force from that wall to compress the beam back between the supports overall (see Video 20: Superposition Problem).

EXAMPLE: COMPOSITE BEAM PROBLEM

The concrete column, capped with a rigid steel plate, is loaded axially in compression by a 50 kip force at the center of the column. There are four $\frac{3}{4}$" diameter steel rebar reinforcements cast within the column. If $E_{CR} = 5 \times 10^3$ ksi, and $E_{ST} = 29 \times 10^3$ ksi, find the normal stress in the concrete and the steel as well as the overall shortening of the beam.

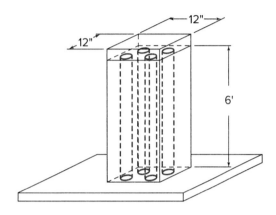

STEP 1: Develop the compatibility equations. Think of compatibility equations as "logic" equations. In this problem, there are two that we need.

1. When the column gets loaded, the concrete will shorten the exact same amount as the steel.

$$\therefore \delta_{ST} = \delta_{CR}$$

2. The 50 kip load will be shared by both the concrete and the steel.

$$\therefore P_{CR} + P_{ST} = 50 \text{ kips}$$

STEP 2: Define the components of the $(\frac{PL}{AE})$ equation for each material and write each in basic units. (Here, we will use kips and inches.)

STEEL

$P_{ST} = ?$ (unknown)

$L_{ST} = 72"$

$A_{ST} = 4(\pi)(\frac{0.75}{2})^2 \text{ in}^2 = 1.767 \text{ in}^2$

$E_{ST} = 29{,}000 \dfrac{\text{kips}}{\text{in}^2}$

CONCRETE

$P_{Cr} = ?$ (unknown)

$L_{Cr} = 72"$

$A_{Cr} = (12" \times 12") - 1.767 \text{ in}^2 = 142.23 \text{ in}^2$

$E_{Cr} = 5{,}000 \dfrac{\text{kips}}{\text{in}^2}$

STEP 3: Find the relationship between P_{ST} and P_{Cr} by substituting into compatibility equation #1:

$$\delta_{ST} = \delta_{CRr}$$

$$\frac{P_{ST}(72")}{(1.767 \text{ in}^2)(29{,}000 \frac{\text{kips}}{\text{in}^2})} = \frac{P_{CR}(72")}{(142.23 \text{ in}^2)(5{,}000 \frac{\text{kips}}{\text{in}^2})}$$

$$13.88 \, P_{ST} = P_{CR}$$

Note: From unit analysis, the units for P are kips.

STEP 4: Take the equation from Step 3 and combine with compatibility equation #2. (Two equations, two unknowns!)

1. $13.88 \, P_{ST} = P_{CR}$
2. $P_{CR} + P_{ST} = 50 \text{ kips}$

Substituting equation 1 into 2:

$$13.88 \, P_{ST} + P_{ST} = 50 \text{ kips}$$
$$14.88 \, P_{ST} = 50 \text{ kips}$$
$$P_{ST} = 3.36 \text{ kips}$$

Now plug P_{ST} into equation 2 to solve for P_{CR}

$$P_{CR} = 50 - 3.36 = 46.64 \text{ kips}$$

STEP 5: Compute stress in each material $(\delta = \frac{PL}{AE})$.

$$\sigma_{ST} = \frac{3.36 \text{ kips}}{1.767 \text{ in}^2} = 1.90 \text{ ksi or } 1{,}900 \text{ psi}$$

$$\sigma_{CR} = \frac{46.64 \text{ kips}}{142.23 \text{ in}^2} = 0.328 \text{ ksi or } 328 \text{ psi}$$

STEP 6: Compute the overall shortening of the column. (Since both materials compress the same amount, we only need to calculate for one of the materials.)

$$\delta_{ST} = \delta_{CR} = \frac{P_{ST}(L_{ST})}{A_{ST}(E_{ST})} = \frac{(3.36 \text{ kips})(72")}{(1.767 \text{ in}^2)(29{,}000 \frac{\text{kips}}{\text{in}^2})} = 0.00472 \text{ in}$$

AXIAL STRESS AND PL/AE

▶ Press pause on video lesson 19 once you get to the workout problem. Only press play if you get stuck.

Find the normal stress in the concrete and in each bar. High strength concrete has a Young's Modulus of 4,200 ksi, and the A992 steel rebar has a Young's Modulus of 29,000 ksi.

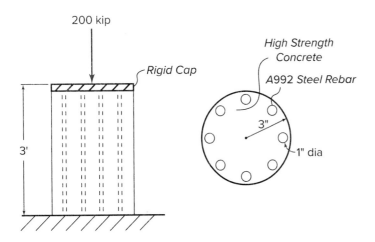

SOLUTION
TO TEST
YOURSELF:
Axial Load
(PL/AE)

✓ TEST YOURSELF 4.3

The assembly below is composed of a solid rod with a diameter of 75 mm, surrounded by a sleeve with an outside diameter of 150 mm and a wall thickness 12.5 mm. Both are held fixed at the wall at left and attached to a rigid plate at right. A force of $F = 50$ kN is applied centrally to the rigid plate. The elastic modulus of the rod is 200 GPa and the elastic modulus for the sleeve is 70 GPa. How much does the assembly elongate due to this load?

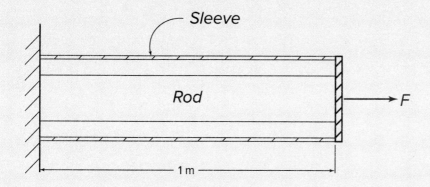

AXIAL STRESS AND PL/AE

 **MECHANICS OF MATERIALS LESSON 20**
Statically Indeterminate Superposition Material Between Two Walls

The following example shows the expansion of materials between two walls due to either a temperature differential or an applied load. The compatibility equation is that the total deflection must be equal to zero.

EXAMPLE: SUPERPOSITION INDETERMINATE

A solid copper rod is bonded to a solid aluminum rod that is constrained between two walls.

- Allowable stress for aluminum is $\sigma_Y = 160$ MPa
- Allowable stress for copper is $\sigma_Y = 110$ MPa

Find the allowable load P that can be applied at point B such that neither bar permanently deforms. Also find the downward displacement of point B.

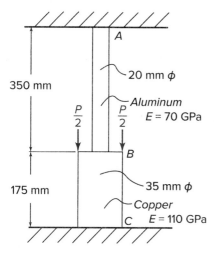

STEP 1:   Ignore one of the supports (in this case, the bottom support). Draw the free body diagram of the system without the bottom support, then sketch in the exaggerated displacement. . . . ♫ ♫ Let it grow, let it grow. ♫ ♫

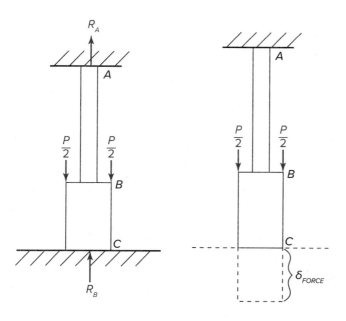

STEP 2: Write the equilibrium equation from FBD #1.

$$\Sigma F_y = 0 = R_A + R_B - P/2 - P/2$$
$$P = R_A + R_B$$

STEP 3: Calculate the total elongation as if the bottom support wasn't there. Understand that during this initial part of the problem, only part AB will be under load.

Hint: Think of a rubber band. If I stretch it from the middle, what's the bottom part of the rubberband experiencing? Nothing.

$$\delta_{AB} = \frac{P(350 \text{ mm})}{\pi(10 \text{ mm})^2(70,000 \frac{N}{mm^2})}$$

$$\delta_{AB} = 1.59 \times 10^{-5}P = \delta_{FORCE}$$

Note: Units for P will be Newtons (N).

STEP 4: Now that we let it stretch (due to force P) we have to push it back so it's between the walls. Which force will push it back into compliance? The reaction force R_B.

Note: When pushing the system back between the two walls, R_B will be the force acting on both parts BC and AB.

$$\delta_{FORCE} = 1.59 \times 10^{-5}P = \delta_{AB} + \delta_{BC}$$

$$1.59 \times 10^{-5}P = \frac{R_B(350 \text{ mm})}{\pi(10 \text{ mm})^2(70,000 \frac{N}{mm^2})} + \frac{R_B(175 \text{ mm})}{\pi(17.5 \text{ mm})^2(110,000 \frac{N}{mm^2})}$$

$$1.59 \times 10^{-5}P = 1.59 \times 10^{-5}R_B + 1.65 \times 10^{-6}R_B$$
$$1.59 \times 10^{-5}P = 1.755 \times 10^{-5}R_B$$
$$R_B = 0.906 \text{ P}$$

STEP 5: Use the allowable stresses to calculate which material will be limiting the load P that can be applied.

P_{MAX} for aluminum: $\quad 160 \text{ MPa} = \frac{P_{AL}}{\pi(10)^2 mm^2}$

$\quad P_{AL} = 50,265.5 \text{ N}$

P_{MAX} for copper: $\quad 110 \text{ MPa} = \frac{P_{CU}}{\pi(17.5)^2 mm^2}$

$\quad P_{CU} = 105,832.4 \text{ N}$

Note: This is the largest force that these materials can stand before permanent deformation. Pick the smallest of these two as it is the one that will cause permanent deformation first. In our problem, the load P_{MAX} is indeed R_B (the compressive force from the lower wall), so we will use 50,265.5 N for R_B.

STEP 6: Substitute $50{,}265.5 \text{ N} = R_B$ into the equation from step 4.

$0.906\,P = 50{,}265.5 \text{ N}$

$P = 55{,}480 \text{ N}$

STEP 7: Finally, compute the displacement of point B downward from the 55.48 kN load. What force is causing tension in Part AB? Answer: R_A.

Let's calculate R_A from our first equilibrium equation.

$R_A + R_B = P \qquad R_A = P - R_B \qquad R_A = 55{,}480 \text{ N} - 50{,}265.5 \text{ N}$

$R_A = 5{,}214.5 \text{ N}$

STEP 8: Calculate δ_{AB} using R_A for the force in section AB.

$$\delta_{AB} = \frac{(5{,}214.5 \text{ N})(350 \text{ mm})}{\pi(10 \text{ mm})^2(70{,}000\,\frac{\text{N}}{\text{mm}^2})}$$

$\delta_{AB} = 0.083 \text{ mm downward}$

▶ Press pause on video lesson 20 once you get to the workout problem. Only press play if you get stuck.

Find the reactions at *A* and *D* from the 200 kN force. The bars are both composed of A-36 steel with a Young's Modulus of 200 GPa.

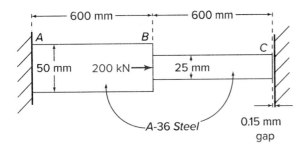

TEST YOURSELF 4.4

SOLUTION TO TEST YOURSELF: Axial Load (PL/AE)

Three cables support a 1,000 lb load as shown below. Each cable has a diameter of ¼ in and a length of 5 ft.

1. What are the stresses in each cable?
2. How far downward does point A move if the cables are made from steel with $E = 29 \times 10^6$ psi?

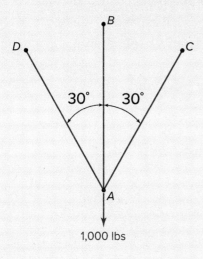

AXIAL STRESS AND PL/AE

 **WATCH VIDEO** **MECHANICS OF MATERIALS LESSON 21**
Thermal Coefficient of Expansion, Axial Elongation

Axial Elongation Due to Temperature Changes

We know from chemistry or material science that when materials experience temperature changes, they will generally expand if the temperature increases (positive Δ) or contract when temperatures decrease (negative Δ).

Materials contract or expand according to their observed material property called the "thermal coefficient of expansion" (α). Also sometimes called the coefficient of thermal expansion (CTE).

The coefficient (α) has strange units of /°C (per degree Celsius) for SI and /°F (per degree Fahrenheit) for imperial units.

We can calculate changes in length in specimens using the following equation:

$$\delta_{TEMP} = \alpha\,(L_O)(\Delta T)$$

where: α = thermal coefficient of expansion (can be found in material property tables)

L_O = original length of the specimen

ΔT = change in temperature (can be positive or negative)

PRO TIP

The challenge you really need to practice is when we combine changes in length due to force (δ_F) with a change in temperature (δ_T) in the same problem. Sounds like a good exam question to me!

EXAMPLE: THERMAL COEFFICIENT OF EXPANSION

A composite bar made from red brass and copper rests perfectly between two walls with zero force in it until it experiences a force of 200 kN as well as a change of temperature from an initial temp of 50° with an increase to 75°. Find the normal stress in each section of the bar and find the displacement of point *B*.

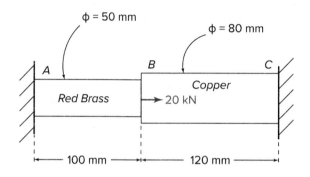

$E_{CU} = 110$ GPa

$\alpha_{CU} = 17.0 \times 10^{-6}/°C$

$E_{BR} = 101$ GPa

$\alpha_{BR} = 18.7 \times 10^{-6}/°C$

STEP 1: We will solve this problem using the "method of superposition." First take one of the support walls away and "let it grow, let it grow." 🎵 🎵

First due to the force (remember the force only affects section *AB*), draw the "exaggerated" growth.

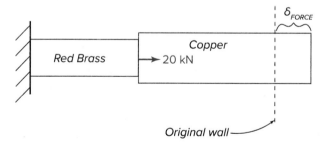

STEP 2: Calculate δ_{FORCE} caused by the 200 kN force (only applies to section *AB*; section *BC* is just "along for the ride").

$$\delta_{FORCE} = \frac{PL}{AE} = \frac{(20,000 \text{ N})(100 \text{ mm})}{\pi(25)^2 \text{ mm}^2(101,000\frac{\text{N}}{\text{mm}^2})} = 0.0101 \text{ mm}$$

STEP 3: Let the bar continue to grow from temperature. The increase in temperature will affect **both** parts of the bar.

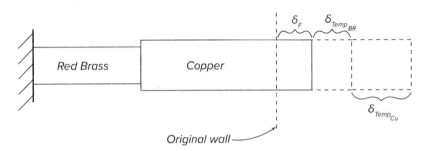

Original wall

STEP 4: Calculate growth from temperature for each section of the bar.

$$\delta_{TEMP_{Total}} = \delta_{TEMP_{BR}} + \delta_{TEMP_{CU}}$$

$$= (18.7 \times 10^{-6}/°C)(100 \text{ mm})(75°C - 50°C) + (17.0 \times 10^{-6}/°C)(120 \text{ mm})(75°C - 50°C)$$

$$= 0.0468 + 0.051 \text{ mm} = 0.0978 \text{ mm}$$

STEP 5: Calculate the total growth of the bar.

$$\delta_{TOTAL} = \delta_{TEMP} + \delta_{FORCE} = 0.0978 + 0.0101 = 0.1079 \text{ mm}$$

STEP 6: Draw a FBD of the original system and write the equation of equilibrium.

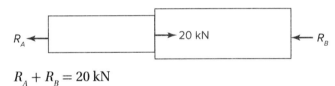

$$R_A + R_B = 20 \text{ kN}$$

STEP 7: Use the reaction at B (R_B) to "push" the newly expanded bar back between the walls. Remember the effect from the 20 kN force and the temperature.

$$\delta_{TOTAL} = \delta_{CU_{R_B}} + \delta_{BR_{R_B}} \Rightarrow$$

$$0.1079 \text{ mm} = \frac{R_C(100 \text{ mm})}{\pi(25)^2 \text{ mm}^2(101,000\frac{N}{mm^2})} + \frac{R_C(120 \text{ mm})}{\pi(40)^2 \text{ mm}^2(110,000\frac{N}{mm^2})}$$

$$0.1079 \text{ mm} = 5.04 \times 10^{-7}R_C + 2.17 \times 10^{-7}R_C$$

$$0.1079 \text{ mm} = 7.21 \times 10^{-7}R_C$$

$$R_C = 149,653 \text{ N or } 149.7 \text{ kN}$$

STEP 8: Find the normal stress in sections AB and BC.

$$\sigma_{AB} = \frac{F_{AB}}{A_{AB}} = \frac{149,700 \text{ N}}{\pi(25)^2 mm^2} = 76.24 \text{ MPa}$$

$$\sigma_{BC} = \frac{F_{BC}}{A_{BC}} = \frac{149,700 \text{ N}}{\pi(40)^2 mm^2} = 29.78 \text{ MPa}$$

▶ Press pause on video lesson 21 once you get to the workout problem. Only press play if you get stuck.

Find the normal stress in each segment of the bar if $T_1 = 70°F$ and $T_2 = 100°F$.

$\alpha_{alum} = 12.8 \times 10^{-6}/°F$ $E_{alum} = 10,600$ ksi

$\alpha_{bronze} = 9.60 \times 10^{-6}/°F$ $E_{bronze} = 5,000$ ksi

$\alpha_{304\ stainless} = 9.60 \times 10^{-6}/°F$ $E_{304\ stainless} = 28,000$ ksi

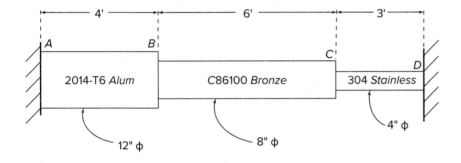

✓ TEST YOURSELF 4.5

SOLUTION TO TEST YOURSELF: Axial Load (PL/AE)

The composite square bar is firmly attached to the two walls and is composed of three firmly attached blocks, each with a cross-sectional area of 1 in^2.

If the temperature drops from 95°F to 25°F, what are the thermally induced stresses in the blocks? For steel, take $E = 29 \times 10^6$ psi and $\alpha = 6 \times 10^{-6}$/°F, and for aluminum take $E = 10 \times 10^6$ psi and $\alpha = 12 \times 10^{-6}$/°F.

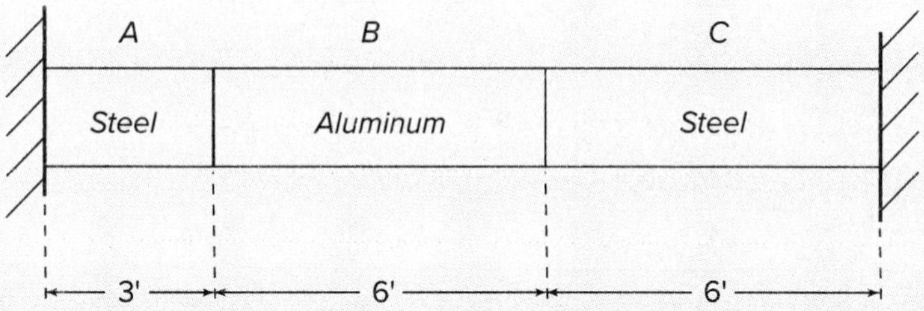

ANSWER

9,688.5 psi (tension)

 **MECHANICS OF MATERIALS LESSON 22**
Stress Riser Concentration Problem; Stress Flow

What Are Stress Concentrations?

This section will discuss stress concentrations, also called "stress risers," that result from axial loads and abrupt changes in geometry.

Think about how stress travels through members under axial loads, called "stress flow," like water in a fast-running river. If we add smaller boulders behind the big one, turbulence is reduced and stress is smoothed out.

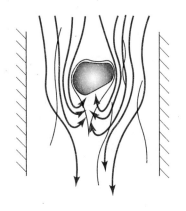
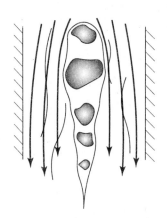

The walls of your house have to carry the load of the roof, attic, and so forth. As that load "flows" through the wall, sometimes it comes to windows or doors with square corners and there arise "stress concentrations" at those corners. That is why you see cracks in mortar or cracks in walls emanating from those corners.

The Roman architects had this figured out over a thousand years ago. Stone buildings had arched or peaked door and window openings to reduce these stress risers.

Finally, here is a computer model, simulating stress in an axially loaded plate with a hole in the center. It may be counterintuitive, but adding *more holes* to this plate actually lowers the stress in the plate in this example by almost 9% by smoothing out the stress flow.

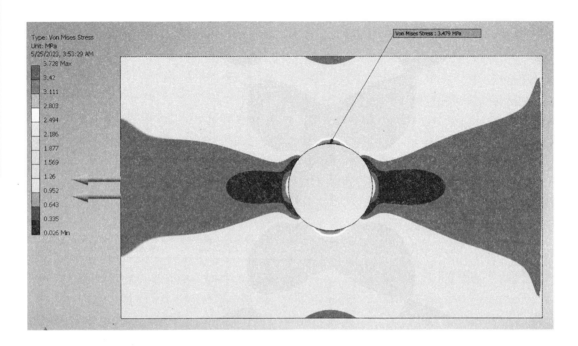

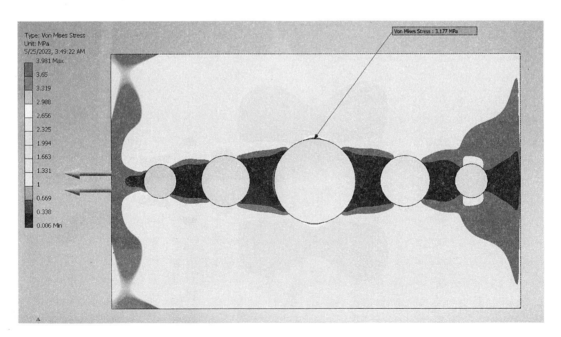

In engineering design, there are typically two types of abrupt geometry changes we are interested in:

1. Sharp corners—typically we reduce stress from sharp corners by adding fillets (radiused corners)
2. Holes

These factors are responsible for a rise in stress that is actually more than $\delta = \dfrac{PL}{AE}$ stress. This increased stress is called σ_{MAX}. The equation to find σ_{MAX} is:

$\sigma_{MAX} = K(\sigma_{nominal})$

$\sigma_{nominal} = $ (normal stress across the smallest cross section)

$\sigma_{MAX} = $ maximum stress due to stress concentration from change in geometry

$K = $ stress concentration factor

The stress concentration factor is computed from the following diagrams.

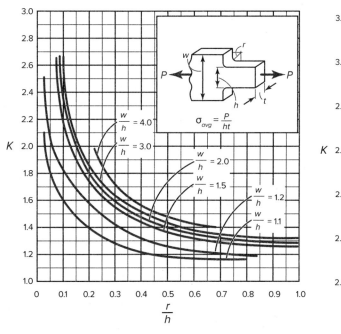

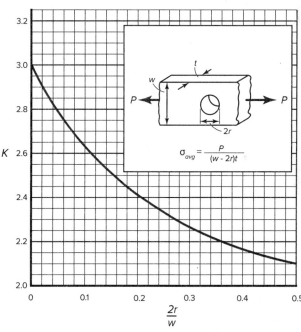

PITFALL Don't get the charts confused. Some of them are for bending stress and others are for an axial load. Pay attention to your loading scenario!

Generally, if you are asked a question about stress concentration, you will be given these charts. The good news is that all of the equations needed are on the diagrams.

EXAMPLE: STRESS CONCENTRATION

The aluminum plate is 12 mm thick. Using a factor of safety of 1.5, find max force P that can be axially applied to the plate. Use $E_{AL} = 70$ GPa and $\sigma_Y = 160$ MPa.

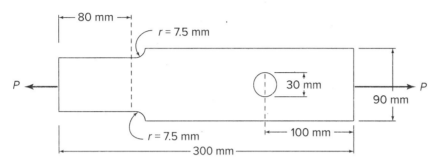

STEP 1: Calculate stress concentration factor (K) for the u-shaped notches.

From chart 1: Find $\dfrac{r}{h} = \dfrac{7.5 \text{ mm}}{75 \text{ mm}} = 0.1$

Calculate $\dfrac{w}{h} = \dfrac{90 \text{ mm}}{75 \text{ mm}} = 1.2$

Follow the 0.21 line vertically upward until it intersects the 1.83 line. Then move left horizontally and read the K value from the chart.

$\therefore K = 1.83$

STEP 2: Calculate stress concentration factor (K) for the hole using the second chart.

From chart 2: Find $\dfrac{2r}{w} = \dfrac{30 \text{ mm}}{90 \text{ mm}} = 0.333$

Follow the 0.33 line vertically upward until it intersects the curve. Move left horizontally to find the value of K.

$\therefore K = 2.23$

STEP 3: Apply factor of safety to yield stress (σ_Y) to find our "allowable" stress, which we will use as our σ_{MAX}.

$\sigma_{MAX} = \dfrac{\sigma_1}{F.S.} = \dfrac{160 \text{ MPa}}{1.5} = 106.67 \text{ MPa}$

STEP 4: Find the value of P using the fillet as the limiting or controlling feature.

$$\sigma_{MAX} = K(\sigma_{nominal})$$

$$\sigma_{MAX} = K(\frac{P}{ht}) \leftarrow \text{Get this equation from the fillet chart.}$$

$$106.67 \frac{N}{mm^2} = 1.83 \left(\frac{P}{(75 \text{ mm})(12 \text{ mm})}\right)$$

$$P = 52,461 \text{ N or } 52.5 \text{ kN}$$

STEP 5: Find the value of P using the hole as the limiting or controlling feature.

$$\sigma_{MAX} = K(\sigma_{nominal})$$

$$\sigma_{MAX} = K(\frac{P}{(w-2r)t}) \leftarrow \text{Get this equation from the hole chart.}$$

$$106.67 \frac{N}{mm^2} = 2.23 \left(\frac{P}{(90 \text{ mm} - 30 \text{ mm})(12 \text{ mm})}\right)$$

$$P = 34,440.5 \text{ N or } 34.4 \text{ kN}$$

STEP 6: Finally, compare the P values from steps 4 and 5 and select the smaller, as the larger value will cause a failure.

$$P_{fillet} = 52.5 \text{ kN}$$

$$P_{hole} = 34.4 \text{ kN}$$

$$\therefore P_{MAX} = 34.4 \text{ kN}$$

Press pause on video lesson 22 once you get to the workout problem. Only press play if you get stuck.

Find the max stress in the bar if $P = 2$ kips. *Hint:* Use the charts at the beginning of this lesson.

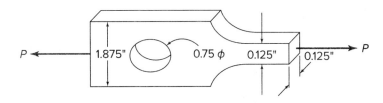

AXIAL STRESS AND PL/AE

☑ **TEST YOURSELF 4.6**

SOLUTION TO TEST YOURSELF: Axial Load (PL/AE)

The bar shown below is subjected to a 10 kip axial load along the length of the bar. Compute the normal stress in the bar (note the ¼" hole is symmetrically drilled) at the stress concentration caused by the small hole.

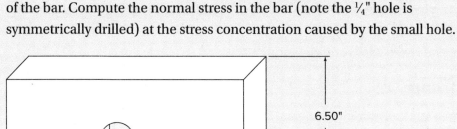

6.50"

1.20"

ANSWER

$\sigma = 3{,}772$ psi

Axial Elongation with Load

- Axial elongation due to a load applied utilizing the "play" equation to determine the change in the axial dimension.
- Elongation is tension, and shrinkage is compression.

Statically Indeterminate

- A problem is considered statically indeterminate if the number of unknowns exceeds the number of equations from static equilibrium.
- Utilizes some type of compatibility equation. Common compatibility equations include the following:
  - **Dependent members.** Equations are visually derived from similar triangle relationships.
  - **Composite materials problems.** Systems composed of multiple materials undergoing an axial load where the change in length of both materials is the same.
  - **Specimen between two walls or supports and a wall.** Utilizes a superposition technique where we ignore the wall and then use the reaction force from that wall to compress the beam back between the supports overall.

Axial Elongation with Temperature Change

- Axial elongation due to a change in temperature utilizes the coefficient of thermal expansion to calculate the change in length of a material.
- Elongation results from heating a material, and compression results from cooling a material.

Stress Risers/Concentration

- Stress risers occur when there are abrupt changes in part geometry. These changes can include holes, U-shaped reliefs, and radiused or square corners.

Equations Learned in this Level

Axial Elongation Due to Load

$$\delta = \frac{PL}{AE}$$

Axial Elongation Due to Change in Temperature

$$\delta_{TEMP} = \alpha (L_O)(\Delta_T)$$

Max Stress Using Stress Concentration Factor

$$\sigma_{MAX} = K(\sigma_{nominal})$$

PRO TIPS

Axial Elongation with Loads
- This is a bit of a repeat, and probably not the last. When using the "play" equation, always include your units in your work as each term in the equation will present you with an opportunity for mixed units! Break your units down to basic units like N and mm for metric or kips and inches for Imperial units.

Axial Elongation with Temp
- The challenge you really need to practice is when we combine changes in length due to force (δ_F) with a change in temperature (δ_T) in the same problem. Sounds like a good exam question to me!

PITFALLS

Axial Elongation with Loads

- A common mistake made when using the "play" equation is which length to use for the L in the equation. It is never the stretched or elongated length, but *always* the original length (L_0) of the specimen!

  Make sure to plug in the proper "sign" of the force: positive if in tension and negative if in compression. This is extremely critical if your shaft has multiple forces being applied at each section.

Stress Risers/Concentration

- Don't get the charts confused. Some of them are for bending stress and others are for an axial load. Pay attention to your loading scenario!

  Generally, if you are asked a question about stress concentration, you will be given these charts. The good news is that all of the equations needed are on the diagrams.

Level 5
Torsion

MECHANICS OF MATERIALS LESSON 23
Shear Stress Due to Torsion, Polar Moment of Inertia

What Is Shear Stress Due to Torsion?

This level will discuss the effects of torsion on members. Typically, we see torsional loadings from power transmission—electric motors turning a shaft, a driveshaft in a car, axles driving wheels, or shafts turning belts on pulleys or chains on sprockets.

- Applying torque (torsion) to a shaft causes shear stress (τ) in that shaft.
- Imagine a torsion spring in a garage door as a solid beam. As a torsional load is applied to a shaft, think of it as layers of material, like the coils of the torsion spring. As torque is applied, the layers slip, one relative to the next, causing shear. If you draw a square element on the shaft, it will turn into a parallelogram.

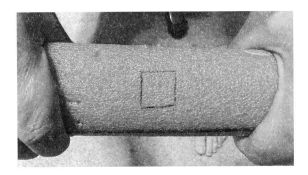

Original stress square   Deformed stress square after torque applied

- Shear stress in shafts caused by torsion is calculated using the following equation:

$$\tau = \frac{Tc}{J}$$

τ = Shear stress

T = Torque

c = Distance from the neutral axis to the point of interest

J = Polar Moment of Inertia

Note: We use "c" instead of "r" as there may be instances where we are interested in finding the shear stress at a particular location in the shaft—not necessarily at the outside diameter only; "c" is used as the distance from the neutral axis to the point of interest.

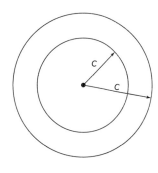

- The Polar Moment of Inertia (J) is the measure of an object's ability to resist torsion or torque on a given axis. Units for Polar Moment of Inertia (J) are the same as the Second Area Moment of Inertia (I): in⁴ or mm⁴.

$$J = \frac{\pi}{2}(r)^4 \text{ for solid} \qquad J = \frac{\pi}{2}\left(r_{outer}^{\;4} - r_{inner}^{\;4}\right) \text{ for hollow}$$

PITFALL

Beware: It is very easy to get the equations for *J* and *I* mixed up. Here's a trick to remember—the line that forms the *J* has two ends, and *I* has four ends.

$$J = \frac{\pi}{2}(r)^4 \qquad I = \frac{\pi}{4}(r)^4$$

Maximum Shear Stress

Maximum shear stress *always* occurs at the outer diameter of shafts. Shear stress is zero at the center of shafts since the distance $c = 0$ (For the elastic region only).

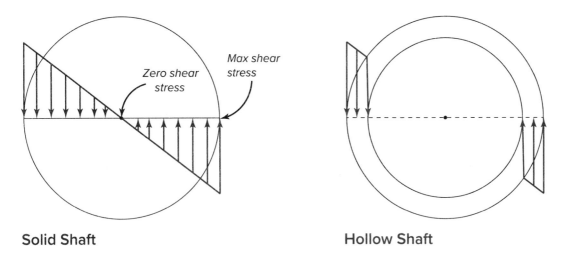

Solid Shaft   **Hollow Shaft**

Another commonly asked question is what is the sign we use for shear stress due to torsion? To determine the sign, simply follow the right hand rule. If the shear stress on the *x* face rotates the *stress element* counterclockwise then the shear stress is positive. If the shear stress on the *x* face rotates the *stress element* clockwise, the shear stress in negative. *Note:* This is assuming your *xy* coordinate axis is oriented as shown in the image.

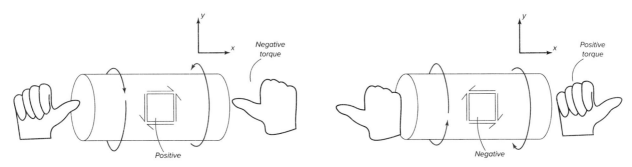

Positive Shear Stress   **Negative Shear Stress**

EXAMPLE: TORSIONAL SHEAR STRESS

The 1.25" diameter solid shaft has the given torques applied. Find the maximum shear stress in the shaft.

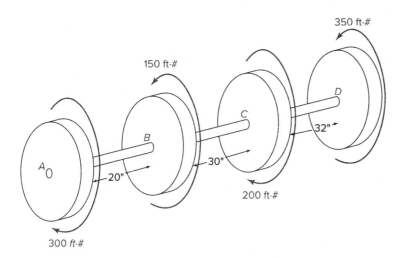

STEP 1: Find the maximum torque by finding the torque in each section. To find the torque, cut the beam and draw the FBD of the easiest side.

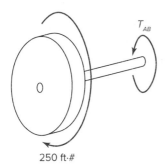

Section *AB*: ΣM (sum of torques) $= 0$
$T_{AB} - 250 \text{ ft·lb} = 0$
$T_{AB} = 250 \text{ ft·lb}$

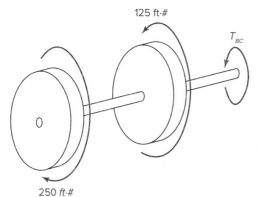

Section *BC*: ΣM (sum of torques) $= 0$
$T_{BC} - 250 \text{ ft·lb} + 125 \text{ ft·lb} = 0$
$T_{BC} = 125 \text{ ft·lb}$

TORSION

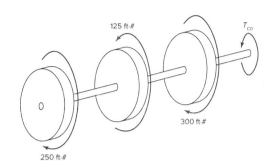

Section *CD*: ΣM (sum of torques) $= 0$

$T_{CD} - 250$ ft·lb $+ 125$ ft·lb $- 300$ ft·lb $= 0$

$T_{CD} = 425$ ft·lb

Therefore: The section with the maximum torque is section *CD*.

STEP 2: Calculate shear stress in section *CD*.

$$\tau_{CD} = \frac{Tc}{J} = \frac{(425 \text{ ft·lb})(\frac{12 \text{ in}}{\text{ft}})(0.625 \text{ in})}{\frac{\pi}{2}(0.625)^4 \text{in}^4} = 13{,}298.7 \text{ psi or } 13.3 \text{ ksi}$$

TORSION

▶ Press pause on video lesson 23 once you get to the workout problem. Only press play if you get stuck.

Find the shear stress in all sections of the hollow shaft. Also plot the torque in the shaft.

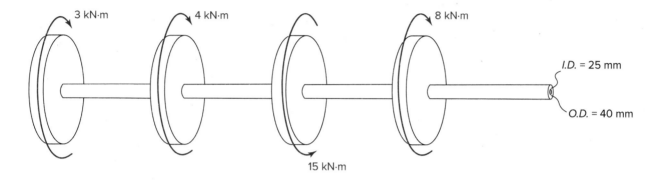

✓ **TEST YOURSELF 5.1**

SOLUTION
TO TEST
YOURSELF:
Torsion

For the three shafts bonded together and loaded as shown, what is the maximum shear stress along the length, and in which section does it occur? Ignore any stress concentrations due to the sudden change in geometry. Take $E = 10 \times 10^6$ psi.

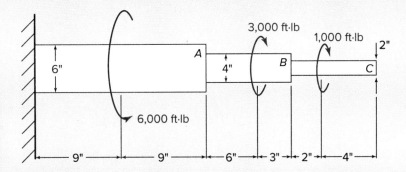

TORSION

MECHANICS OF MATERIALS LESSON 24
Power Transmission Torque Example

Power Transmission in a Shaft

Often we will be given power ratings for rotating machinery. Remember, power is the rate of doing work. For the metric system, power will be given in watts (W) or kilowatts (kW), and for US customary units, horsepower.

- Convenient Conversions (Read Necessary Conversions—you are going to need these!)

$$\text{watt} = \frac{\text{N·m}}{\text{s}} \qquad \text{kW} = \frac{1,000 \text{ N·m}}{\text{s}} \qquad 1\text{hp} = \frac{550 \text{ ft·lb}}{\text{s}}$$

- The equation relating power to torque is as follows:

  $P = T\omega$ $P = $ power

  or $T = $ torque

  $T = \dfrac{P}{\omega}$ $\omega = $ angular velocity in $\dfrac{\text{rad}}{\text{s}}$

- Angular velocity (ω) is commonly given as rpm, but we need ($\dfrac{\text{rad}}{\text{s}}$). This is an easy conversion to make.

$$\frac{\text{rev}}{\text{min}} \times \frac{2\pi \text{ rad}}{\text{rev}} \times \frac{\text{min}}{60\text{s}} = \frac{\text{rad}}{\text{min}}$$

- Example for conversion of 500 rpm to $\dfrac{\text{rad}}{\text{s}}$:

$$500 \times \frac{2\pi}{60} = 52.36 \, \frac{\text{rad}}{\text{s}}$$

- Commonly, we will be required to use power and angular velocity to calculate torque (T). Once the T is calculated, the shear stress can be determined.

PITFALL

Beware: It is very easy to forget that the angular velocity (ω) has to be in rad/s and not rpm!

TORSION

EXAMPLE: POWER TRANSMISSION TORQUE

The given system has an electric motor running at 3,000 rpm. It delivers 3 kW and 8 kW to two systems. Determine the minimum diameter of the solid steel output shaft. $\tau_{Allow} = 70$ MPa.

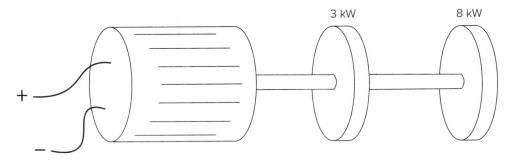

3 kW   8 kW

STEP 1: Determine total power.

$$3 \text{ kW} + 8 \text{ kW} = 11 \text{ kW} = 11,000\frac{\text{N·m}}{\text{s}}$$

STEP 2: Convert rpm to $\frac{\text{rad}}{\text{s}}$.

$$\omega = 3,000 \text{ rpm} \times \frac{2\pi}{60} = 314.2\frac{\text{rad}}{\text{s}}$$

STEP 3: Use the power equation to derive the torque required for the system.

$$T = \frac{P}{\omega}$$

$$T = \frac{11,000\dfrac{\text{N·m}}{\text{s}}}{314.2\dfrac{\text{rad}}{\text{s}}} = 35.01 \text{ N·m}$$

STEP 4: Use the shear stress equation to find the radius (and then the diameter) of the shaft.

$$\tau = \frac{Tc}{J} \qquad 70\frac{\text{N}}{\text{mm}^2} = \frac{35.01 \text{ N·m}\dfrac{(1,000 \text{ mm})}{\text{m}}(r)\text{mm}}{\dfrac{\pi}{2}(r^4)\text{mm}^4} \qquad 70r^3 = \frac{35,010}{\dfrac{\pi}{2}}$$

$$r^3 = 318.4 \qquad r = 6.83 \text{ mm} \qquad \text{diameter} = 13.66 \text{ mm}$$

TORSION

Press pause on video lesson 24 once you get to the workout problem. Only press play if you get stuck.

Find the shear stress in section *AB* and *BC* of the 25 mm diameter solid shaft. The motor is running at 50 rev/s.

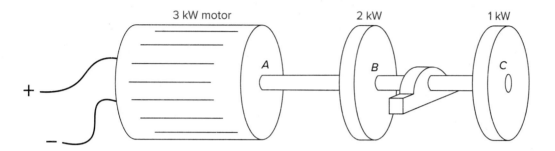

✓ TEST YOURSELF 5.2

SOLUTION TO TEST YOURSELF: Torsion

A turbine engine produces 7,000 kW at 5,500 RPM. A solid circular shaft of diameter D_1 connects the output to a 10:1 reduction gear set. From there, a second solid shaft of diameter D_2 connects the electrical generator.

Note: 10:1 signifies that shaft 2 is rotating at 10 times less than shaft 1 (i.e., its angular velocity [ω] is less).

If the allowable shear stress in the shaft material is 100 MPa, what are the diameters of the two solid circular shafts?

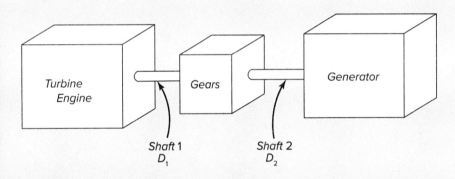

Turbine Engine

Gears

Generator

Shaft 1
D_1

Shaft 2
D_2

TORSION

ANSWERS

$D_1 = 85.2$ mm

$D_2 = 183.6$ mm

MECHANICS OF MATERIALS LESSON 25
Angle of Twist Due to Torque, Torsion

Angle of Twist

As torque or torsion is applied to any shaft, a twist will be induced into the material. This is called the angle of twist, and this angle is always given in radians unless specified otherwise. The angle of twist of any shaft is as follows:

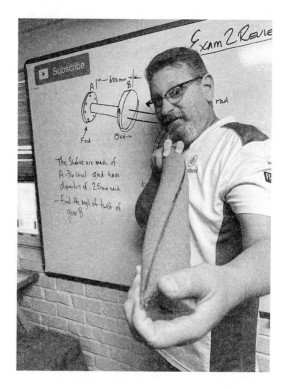

$$\phi = \frac{TL}{JG}$$

ϕ = angle of twist in radians!

T = torque

L = shaft length

J = polar moment of inertia

G = shear modulus of rigidity

Angle of twist is calculated as radians, which means that on the right side of the equation, *all units should cancel out.*

There are lots of opportunities for mixed units, so make sure to reduce everything to basic units like N and mm or kips and inches.

The easiest way to calculate the angle of twist is to consider one end of the shaft to be fixed, and then determine the twist in the rest of the shaft relative to that fixed end.

PRO TIP

To help visualize the angle of twist, get yourself a pool noodle (split it into thirds and share with two classmates), and with a marker, draw a straight line down the length of the noodle. Now you have a model you can twist and have a visual representation of what is going on in your homework problem as you hold one end and twist the other!

EXAMPLE: ANGLE OF TWIST

Find the total angle of twist of end *D* with respect to end *A*. The shafts are all 1" in diameter and take $G = 11{,}000$ ksi.

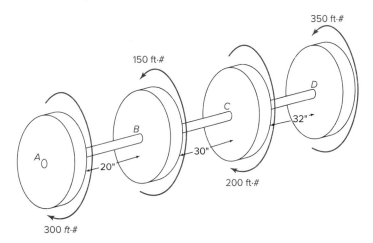

STEP 1: Determine the torque in each section of the shaft. Also determine the direction according to the right-hand rule.

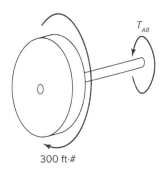

Section *AB*: ΣM (sum of torques) $= 0$

$T_{AB} - 300 \text{ ft·lb} = 0$

$T_{AB} = 300 \text{ ft·lb}$

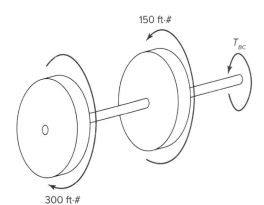

Section *BC*: ΣM (sum of torques) $= 0$

$T_{BC} - 300 \text{ ft·lb} + 150 \text{ ft·lb} = 0$

$T_{BC} = 150 \text{ ft·lb}$

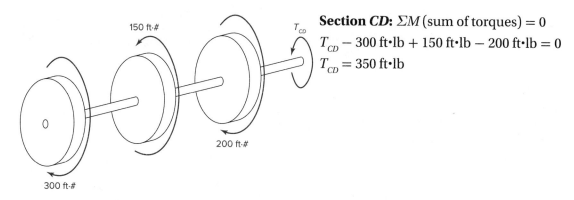

Section *CD*: ΣM (sum of torques) $= 0$

$T_{CD} - 300 \text{ ft·lb} + 150 \text{ ft·lb} - 200 \text{ ft·lb} = 0$

$T_{CD} = 350 \text{ ft·lb}$

STEP 2: Calculate the total angle of twist using the following equations:

$$\phi_{AD} = \phi_{AB} + \phi_{BC} + \phi_{CD}$$

$$\phi_{AD} = \frac{(T_{AB})(L_{AB})}{JG} + \frac{(T_{BC})(L_{BC})}{JG} + \frac{(T_{CD})(L_{CD})}{JG}$$

$$\phi_{AD} = \frac{(300 \text{ ft·lb})(24 \text{ in})}{\dfrac{\pi}{2}(0.5)^4 \text{ in}^4 \left(11{,}000 \dfrac{\text{kips}}{\text{in}^2} \times \dfrac{1{,}000 \text{ lb}}{\text{kip}}\right)} + \frac{(150 \text{ ft·lb})(30 \text{in})}{\dfrac{\pi}{2}(0.5)^4 \text{ in}^4 \left(11{,}000 \dfrac{\text{kips}}{\text{in}^2} \times \dfrac{1{,}000 \text{ lb}}{\text{kip}}\right)}$$

$$+ \frac{(350 \text{ ft·lb})(32 \text{ in})}{\dfrac{\pi}{2}(0.5)^4 \text{ in}^4 \left(11{,}000 \dfrac{\text{kips}}{\text{in}^2} \times \dfrac{1{,}000 \text{ lb}}{\text{kip}}\right)}$$

$$\phi_{AD} = 0.00667 + 0.00417 + 0.01037$$

$$\phi_{AD} = 0.0212 \text{ radians clockwise or } 0.0212 \text{ radians} \times \frac{360°}{2\pi \text{ radians}} = 1.215°$$

▶ Press pause on video lesson 25 once you get to the workout problem. Only press play if you get stuck.

Find the angle of twist of the shaft of point *F* with respect to point *A*. The shaft is A-36 steel with a shear modulus of 75 GPa.

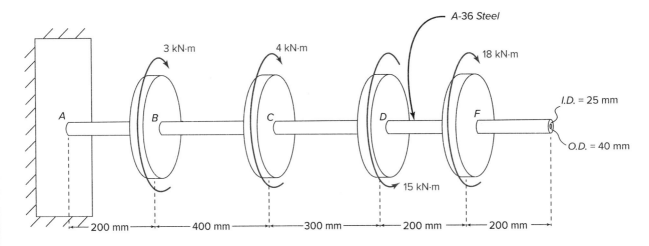

SOLUTION TO TEST YOURSELF: Torsion

✓ TEST YOURSELF 5.3

Determine the horsepower (hp) transmitted by a solid shaft rotating at 150 rpm with diameter of 10 in and a length of 8.5 ft if the observed twist angle of the shaft is 2° over the full shaft length. What is the maximum shear stress? Take $G = 15,000$ ksi.

ANSWERS
hp = 11,992
$\tau = 25,658$ psi

MECHANICS OF MATERIALS LESSON 26
Statically Indeterminate Angle of Twist Due to Torque

Gear Ratio Relationships

We can also develop equations of rotation, or angles of twist (ϕ) from gear ratios any time two different diameter gears come in contact with each other. We have to determine how to make the following relationship true:

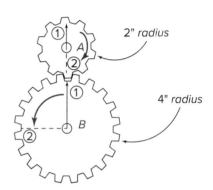

2" radius

4" radius

$s_A = s_B$ the arc length traveled by both gears must be the same

$r_A \phi_A = r_B \phi_B$ relating the rotation of gear A to gear B. **Note:** These are the radii of the gears not the radii of the shafts.

$$\text{Gear Ratio} = \frac{\text{radius of big gear}}{\text{radius of small gear}}$$

or

$$\text{Gear Ratio} = \frac{\text{\# of teeth of big gear}}{\text{\# of teeth of big gear}}$$

If gear A rotates clockwise from position CD to position (2) (180°), then gear B would rotate counterclockwise from position (1) to position (2) (90°). So for the rotation of the two gears to be equal, the following will have to be true:

$$\phi_A = 2\,\phi_B$$

PRO TIP

Whether the gear ratio is 2:1, 3:1, or 7:1, the big number (2, 3, or 7) always goes with the big gear!
Remember this trick as it is very easy to confuse this and write your compatibility equation backward!

TORSION

EXAMPLE: GEAR PROBLEM

Determine the maximum shear stress on shaft *BD*.

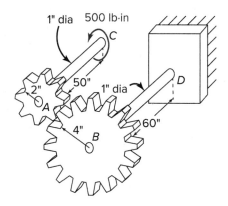

STEP 1: Determine the gear ratio between the two gears in contact with each other (gear *A* and gear *B*).

Note: This can be done by using one of two methods:

1. One diameter (or radius) compared to the other
2. Number of teeth on one gear compared to the number of teeth on the other

For this example:

$r_A = 2"$

$r_B = 4"$

$$\text{Gear Ratio} = \frac{\text{radius of big gear}}{\text{radius of small gear}} = \frac{4"}{2"} = 2$$

∴ the gear ratio is 2:1

STEP 2: Draw a FBD of each shaft separately.

Note: A common mistake is to leave off the contact force between the teeth on the two gears. (*Remember:* The contact forces on each FBD are the same but in opposite directions.)

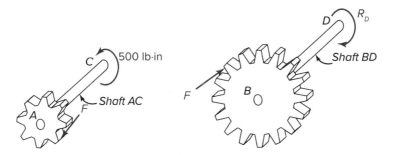

STEP 3: Write equations of equilibrium for each of the shafts (ΣM or sum of the torques).

Shaft AC: $\Sigma M = 0 = 500 \text{ lb·in} - \text{F}(2 \text{ in})$

$$F = \frac{500 \text{ lb}}{2 \text{ in}} = 250 \text{ lb}$$

Shaft BD: $\Sigma M = 0 = \text{F}(4 \text{ in}) - R_D$

$R_D = 250 \text{ lb} \times 4 \text{ in}$

$R_D = 1{,}000 \text{ lb·in}$

STEP 4: Calculate the shear stress in *BD*.

$$\tau_{BD} = \frac{Tc}{J} = \frac{1{,}000 \text{ lb·in } (0.5 \text{ in})}{\frac{\pi}{2}(0.5)^4 \text{in}^4} = 5.092 \times 10^3 \text{ psi}$$

TORSION

TEST YOURSELF 5.4

SOLUTION TO TEST YOURSELF: Torsion

A torque of 175 N-m is applied to 35 mm diameter shaft *A* that is spinning at 500 rpm and connected to the gears as shown. If shaft *B* is to have the same maximum shear stress as shaft *A*, what diameter should be used for shaft *B*?

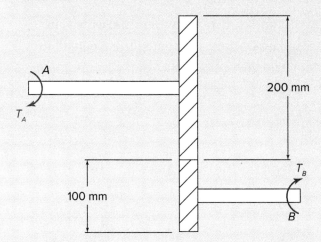

TORSION

Statically Indeterminate Torsion Problems

Very similarly to the last level when we discussed statically indeterminate problems, we will have to write compatibility equations. For these problems, equations typically come from two sources:

1. Global equilibrium and free body diagrams
2. Geometrical constraints (typically gear ratio relationships or angle of twist for torsion problems)

Always start these problems by drawing FBDs of each component of the system and write your equations of equilibrium. To solve problems with more unknowns than your equilibrium equations can handle, we need to generate a compatibility equation and then equation(s) that bring these two together (often called constitutive equation[s]).

PRO TIP

The most common mistake on statically indeterminate problems is to draw an incorrect FBD. I like to think of myself as the "free" body. If I was the gear-shaft system, what would I be feeling?

You are a free body: what is the world doing to you? You are looking for those external forces.

Press pause on video lesson 26 once you get to the workout problem. Only press play if you get stuck.

For the steel shaft, find the largest shear stress in the assembly. Steel has a shear modulus of 10.8×10^3 ksi.

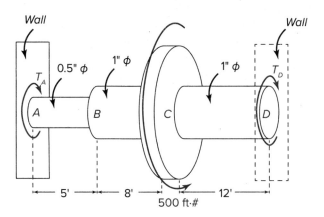

TEST YOURSELF 5.5

SOLUTION TO TEST YOURSELF: Torsion

A solid steel shaft of diameter 5 in and length 10 ft is fastened to rigid plates on both ends. Also attached to the rigid plates is an aluminum sleeve with inner diameter of 12 in and sleeve thickness of 1.5 in. If an equal and opposite torque of 5,000 ft·lb is applied at the ends, determine the maximum shear stress in the shaft and sleeve. Take G_{steel} = 11,000 ksi and $G_{aluminum}$ = 3,800 ksi.

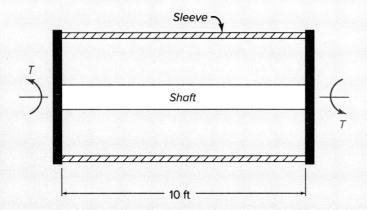

TORSION

MECHANICS OF MATERIALS LESSON 27
Indeterminate Angle of Twist with Gear Ratios

There are three main types of statically indeterminate torsion problems:

1. Materials between walls so that the total angle of twist is zero
2. Sleeved shafts (i.e., a composite material with one material on the outside and a different material for the core)
3. Statically indeterminate gear problems (i.e., a shaft with one end fixed to the wall and the other end with loaded gears)

The first two types were reviewed in the previous lesson. The next calculator problem will showcase how to calculate statically indeterminate problems with gears.

TORSION

EXAMPLE: STATICALLY INDETERMINATE TORQUE

Determine the reactions at *C* and *D*, as well as the angle of twist of gear *A*. $G = 11 \times 10^3$ ksi.

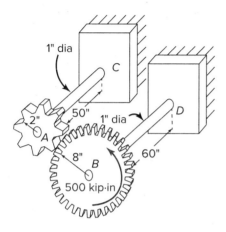

STEP 1: Determine the gear ratio between the two gears in contact with each other (Gear *A* and Gear *B*).

Note: This can be done by using one of two methods:

1. One diameter (or radius) compared to the other
2. Number of teeth on one gear compared to the number of teeth on the other

For this example:

$r_A = 2"$

$r_A = 8"$

$$\text{Gear Ratio} = \frac{\text{radius of big gear}}{\text{radius of small gear}} = \frac{8"}{2"} = 4$$

∴ the gear ratio is 4:1

STEP 2: Draw a FBD of each shaft separately.

Note: A common mistake is to leave off the contact force between the teeth on the two gears.

(*Remember:* The contact forces on each FBD are the same but in opposite directions.)

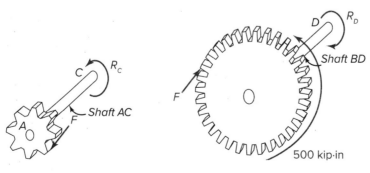

STEP 3: Write equations of equilibrium for each of the shafts (ΣM or sum of the torques).

Shaft *AC*: $\Sigma M = 0 = R_C - F(2")$

$R_C = 2F \Rightarrow F = \dfrac{R_C}{2}$

Shaft *BD*: $\Sigma M = 0 = 500$ kip in $- F(8) - R_D$

$R_D = 500 - 8F$

STEP 4: Substitute equation 1 into equation 2 to get R_C in terms of R_D.

$R_D = 500 - 8\left(\dfrac{R_C}{2}\right)$

$R_D = 500 - 4\,R_C$

STEP 5: Use gear ratio relationship (compatibility) to relate the angles of twist of each shaft. From Step 1, the gear ratio is 4:1; therefore

$4\Phi_B = \Phi_A$

STEP 6: Expand Step 5 with the angle of twist equation and derive a second relationship between R_A and R_D.

$$\dfrac{4(R_D)(L_{BD})}{JG} = \dfrac{(R_A)(L_{AC})}{JG}$$

Since J and G are the same for both shafts, those divide away leaving:

$4(R_D)(L_{BD}) = (R_C)(L_{AC})$

$4\,R_D(60) = R_C(50)$

$4.8\,R_D = R_C$

STEP 7: Substitute Step 6 into the Step 4 equation.

$R_D = 500 - 4\,(4.8\,R_D)$

$R_D = 500 - 19.2\,R_D$

$20.2\,R_D = 500$

$\therefore R_D = 24.75$ in·kip

$R_C = 4.8\,R_D = 4.8\,(24.75)$

$\therefore R_C = 118.8$ in·kip

STEP 8: Finally find the angle of twist of gear *A* by looking at the angle of twist for the shaft *AC*.

$$\Phi_A = \dfrac{TL}{JG} = \dfrac{(R_A)(L_{AC})}{JG} = \dfrac{(118.8 \text{ in·kips})(50 \text{ in})}{\dfrac{\pi}{2}(0.5)^4 \text{in}^4(11{,}000\,\dfrac{\text{kips}}{\text{in}^2})}$$

$\Phi_A = 5.5$ radians

Press pause on video lesson 27 once you get to the workout problem. Only press play if you get stuck.

If 500 Nm is applied to gear E, find the rotation of this gear. The gears and the shafts are constructed of A-36 steel with a shear modulus of 75 GPa.

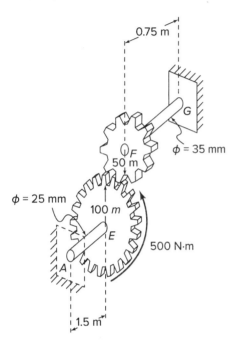

☑ **TEST YOURSELF 5.6**

The solid circular shafts shown below are made from A-36 steel and have the same diameter of 4 in. Find the angle of twist at B. **Note:** The shafts are fixed at A and E.

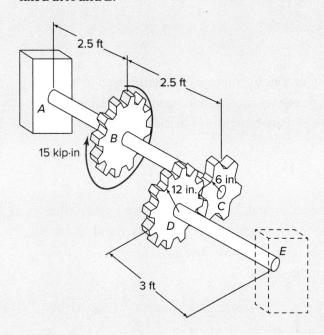

TORSION

Torsional Shear Stress

- When solid circular shafts experience a torsional load, a shear stress is induced in the shaft with the maximum shear stress at the outer fibers of the shaft and zero shear stress in the center of the shaft.

Power Transmission

- Often we will be given the power transmission for rotating machinery. Remember, power is the rate of doing work. For the metric system, power will be given in watts (W) or kilowatts (kW), and for US customary units, horsepower. Remember, the angular velocity needs to be converted to radians per second!

Angle of Twist

- As torque is applied to any shaft, a twist will be induced into the material. This is called the angle of twist, and this angle is always given in radians unless specified otherwise. For a shaft with multiple segments, to determine the total angle, combine the angles in each segment. Remember the rotations could be additive or subtractive.

Statically Indeterminate

- These problems usually show up in one of two forms: a shaft or a stepped shaft between two walls, or two shafts connected by some gears set with some given gear ratio.
- In order to solve these problems, you will have to write your equilibrium equations from free body diagrams, construct compatibility equations by using the geometry of the system, and then write an equation(s) that relates these two (some professors call this constitutive equations). Typically for torsion problems the compatibility equation(s) deal(s) with angle of twists or gear ratios.

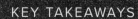

Equations Learned in this Level

Shear Stress Due to Torsion

$$\tau = \frac{Tc}{J}$$

Power Transmission Equation

$$P = T\omega$$

Angle of Twist

$$\phi = \frac{TL}{JG}$$

PRO TIPS

Angle of Twist

- To help visualize the angle of twist, get yourself a pool noodle (split it into thirds and share with two classmates), and with a marker, draw a straight line down the length of the noodle. Now you have a model you can twist and have a visual representation of what is going on in your homework problem as you hold one end and twist the other!

Statically Indeterminate

- Whether the gear ratio is 2:1, 3:1, or 7:1, the big number (2, 3, or 7) always goes with the big gear!

  Remember this trick as it is very easy to confuse this and write your compatibility equation backward!

- The most common mistake on statically indeterminate problems is to draw an incorrect FBD. I like to think of myself as the "free" body. If I was the gear-shaft system, what would I be feeling?

  You are a free body: what is the world doing to you? You are looking for those external forces.

PITFALLS

Torsional Shear Stress

- *Beware:* It is very easy to get the equations for J and I mixed up. Here's a trick to remember—the line that forms the J has two ends, and I has four ends.

$$J = \frac{\pi}{2}(r)^4 \qquad I = \frac{\pi}{4}(r)^4$$

Power Transmission

- *Beware:* It is very easy to forget that the angular velocity (ω) has to be in rad/s and not rpm!

Level 6
Bending

Shear and Moment Diagrams

Flexure Formula

WATCH VIDEO **MECHANICS OF MATERIALS LESSON 28**
Beam Bending, Shear Moment Diagram Example

Things to Remember from Statics to Calculate Bending Stress

- **Centroids.** Used to calculate the location of the neutral axis on a beam's cross section. Use this to find "*c*" in the Flexure Formula (covered later in this level).
- **Global Equilibrium.** Necessary to compute reaction forces on beams (some of these beams are pin-jointed and are solved as frame problems).
- **Shear/Moment Diagrams.** Used to find maximum bending moment and shear force on a loaded beam (M_{max}).

Bending Stress, *M*-calculations

To calculate the bending stress in a beam, you must calculate *M* (bending moment). If the problem asks to solve for the bending stress at a particular point or section in the beam, it is fastest to just "cut" the beam and solve for the internal bending moment. If the problem asks for the maximum bending stress in the entire beam, you will need to figure out where this occurs, which typically requires drawing the entire bending moment diagram.

Shear/Moment Diagram Review

There are two methods to determine the moment diagram:

1. The graphic method
2. The equation method

The graphic method is fastest, so we recommend using that technique.

The Graphic Method for Solving

- Utilizes tables of common shapes (triangles, rectangles, etc.) found in the back of the book.
- Shear diagram (*V*) is the integral of the load curve (the original given loaded beam).
- Moment diagram (*M*) is the integral of the shear curve.

Loading Type	Shear Diagram	Moment Diagram
Point force	Jump up or down equal to the point load, then constant until load changes	Sloped line
Rectangular distributed load	Slope of line equal to the distributed load	Parabolic line
Triangular distributed load	Parabolic line	Cubic line

Coupled moment	No effect	Jumps up or down equal to the coupled moment
		
		When working problem from left to right:
		    CW M is upward jump
		    CCW M is downward jump

Another way to think about the graphic method is as follows:

The Order of the Lines

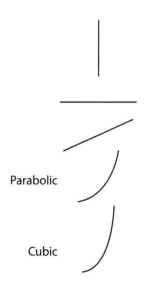

Parabolic

Cubic

- The top line is a concentrated load.
- The next graph down would be horizontal, $y =$ constant.
- If we integrate, we will have an x (linear with constant slope).
- If we integrate again, we will have an x^2 (parabolic).
- If we integrate again, we will have an x^3 (cubic).

Note: If you understand this concept, you should always know what to expect on the next graph as you go from load (L) to shear (V) to the moment (M) graph.

Shear and Moment Diagram—Graphic Method Recipe

STEP 1: Find the reaction forces. Remember that you can convert distributed loads to a concentrated load when solving for reactions, but you must treat them as distributed loads when drawing the *V* and *M* diagram.

STEP 2: Draw the axis of the shear and moment diagram. It is helpful to align the axes.

STEP 3: Put a vertical dashed line on your graph wherever the load changes. These locations mark discontinuities. (Examples include distributed loads changing from a rectangular shape to a triangular shape or a distributed load changing to a concentrated load.)

STEP 4: Draw the shear force diagram and pretend you are a wagon accumulating loads as you move across the beam.

STEP 5: Calculate the "area" under the shear diagram.

STEP 6: If the area on the shear diagram is positive, you add that value to your moment diagram. If it is negative, you subtract that value. The only detail missing is how to deal with concentrated moments. Concentrated moments don't affect the shear diagram, but they will make you jump up or down on the moment diagram.

BENDING

EXAMPLE: SHEAR MOMENT DIAGRAM

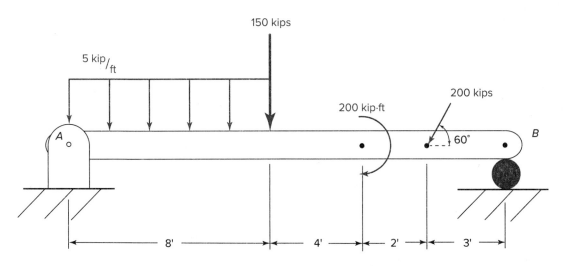

STEP 1: Find global equilibrium.

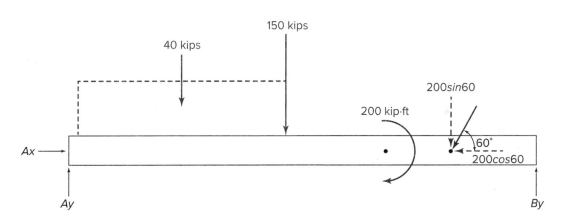

$\Sigma M_A = 0 = -40(4) - 150(8) - 200 - 200sin(60°)(14) + B_y(17)$

$160 + 1{,}200 + 200 + 2{,}424.87 = B_y(17)$

$3{,}984.87 = B_y(17)$

$B_y = 234.4$ kips

$\Sigma F_X = 0 = A_x - 200cos(60°)$

$A_x = 100$ kips

$\Sigma F_Y = 0 = A_Y + 234.4 \text{ kips} - 40 - 150 - 200sin(60°)$

$A_Y = 128.8$ kips

STEP 2: Plot the V diagram using the forces found in global equilibrium. Don't forget the "order of the lines."

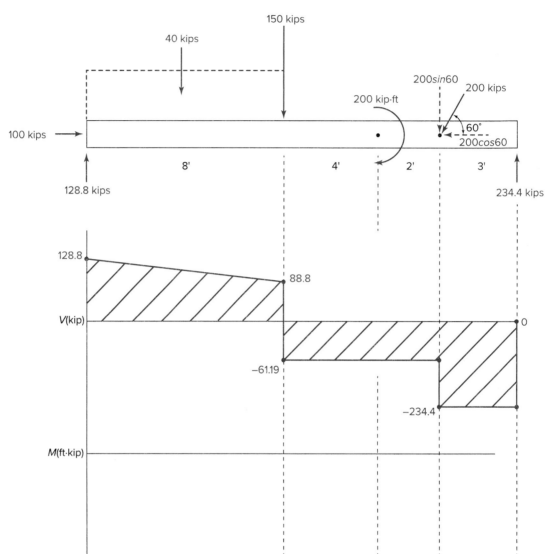

STEP 3: Calculate areas of shapes on the *V* diagram. (Put a circle around them to easily identify.) From there, create your *M* diagram.

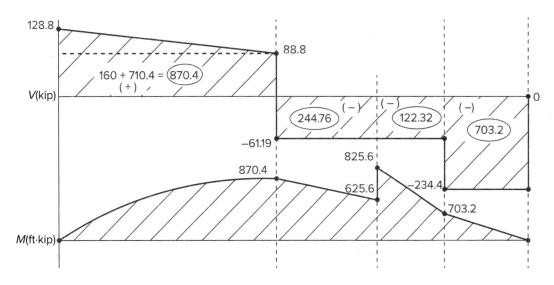

Note: We had to split the middle "rectangle" since there was a concentrated moment right in the middle of it.

Press pause on video lesson 28 once you get to the workout problem. Only press play if you get stuck.

Draw the shear and moment diagram for a beam loaded as follows.

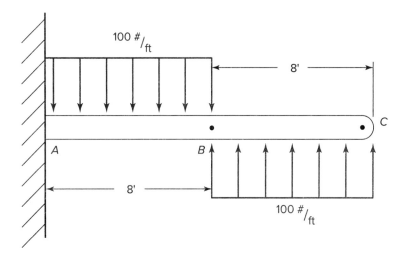

SOLUTION TO TEST YOURSELF: Bending

✓ TEST YOURSELF 6.1

Sketch the shear and moment diagram for the beam shown below.

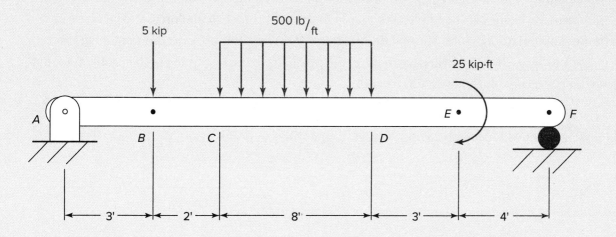

ANSWERS

$M_{max} = 16,040$ ft·lb

$M_{min} = -9,800$ ft·lb

$V_{max} = 5,200$ lb

$V_{min} = -3,800$ lb

MECHANICS OF MATERIALS LESSON 29
Shear Moment Diagram, Graphic Method . . . Challenging

The challenging shear-moment diagram problems come when there is a composite beam (i.e., a pin in the middle of the beam). This complicates your global equilibrium calculation as the beam turns into a frame problem from Statics. The bad news is that if you mess this up, your moment diagram won't go back to zero. Thus, you will have to go all the way back to the beginning of the problem to find out that your global equilibrium was not correct.

▶ Press pause on video lesson 29 once you get to the workout problem. Only press play if you get stuck.

Draw the shear and moment diagram for a beam loaded as follows.

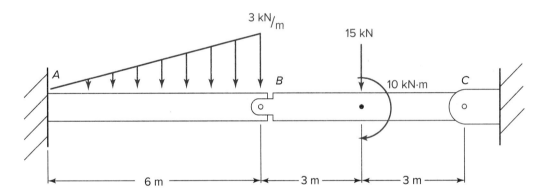

TEST YOURSELF 6.2

SOLUTION TO TEST YOURSELF: Bending

Sketch the shear and moment diagram for the beam shown below.
Note there is a hinge at mid-span.

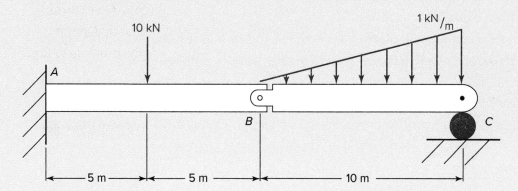

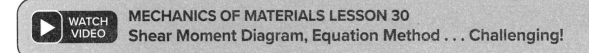

MECHANICS OF MATERIALS LESSON 30
Shear Moment Diagram, Equation Method . . . Challenging!

Shear/Moment Diagrams—The Equation Method

When constructing shear moment diagrams, there may arise a situation where the graphic method of plotting diagrams will not work.

The situation is as follows: the shear diagram is parabolic and crosses the zero axis on the graph.

Note: This can only occur on loaded beams with triangular loads or parabolic loads.

A "hybrid approach" will also work on these problems, where you use the graphic method for all sections except the parabolic section that crosses the axis.

The following example is of a hybrid approach to solving a very difficult shear moment diagram.

EXAMPLE: SHEAR MOMENT DIAGRAM–EQUATION

Draw the shear/moment diagram for the following:

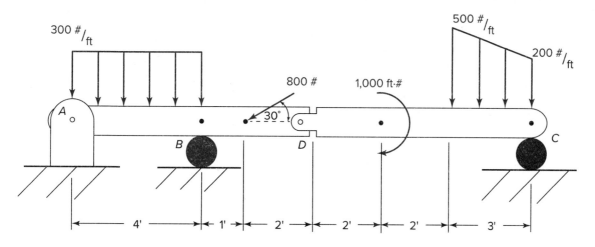

STEP 1: Find global equilibrium. Recognize that this is a frame problem. Draw the FBD of each section of the beam.

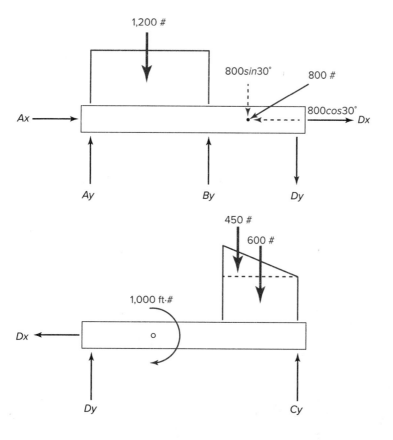

STEP 2: Solve for global equilibrium.

From FBD#2:

$\Sigma M_D = 0 = -1,000 - 450(5) - 600(5.5) + C_y(7)$

$0 = 6,550 = C_y(7)$

$935.7 \text{ \#} = C_y$

$\Sigma F_y = 0 = C_y + D_y - 600 - 450$

$0 = 935.7 + D_y - 1,050$

$D_y = 114.3 \text{ \#}$

Plugging this info into FBD#1:

$\Sigma M_A = 0 = -1,200(2) + B_y(4) - 800(sin(30°))(5) - 114.3(7)$

$0 = -2,400 + 4B_y - 2,000 - 800.1$

$4B_y = 5,200.1$

$B_y = 1,300 \text{ \#}$

$\Sigma F_y = 0 = A_y + 1,300 - 114.3 - 1,200 - 800(sin(30°))$

$A_y = 414.3\text{\#}$

$\Sigma F_D = 0 = A_X - 800(cos(30°))$

$A_X = 692.8 \text{ \#}$

BENDING

STEP 3: Construct the shear force (V) diagram by first adding all discontinuities. Then plot V.

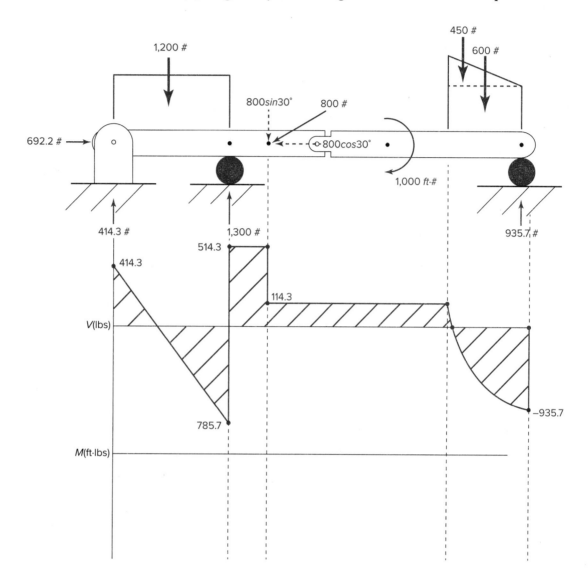

STEP 4: Calculate the areas of all shapes except the parabolic area; then plot the M diagram up to the parabolic section. Draw a circle around areas to keep them separate.

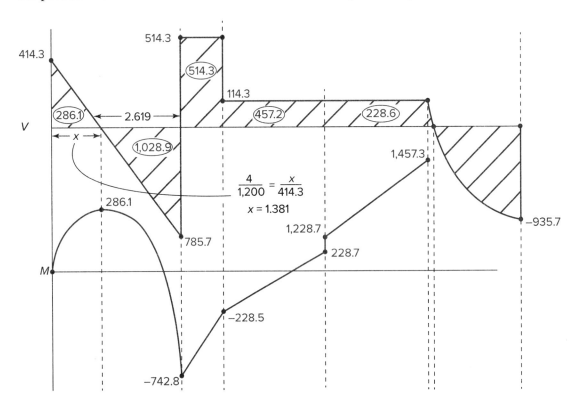

Note: We know the shape of the last two sections:

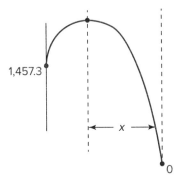

but we don't know what that max value is or exactly where it occurs. This is where the equation method comes in handy.

STEP 5: Cut the original loaded beam on the right-hand side at a distance of x and draw the FBD of that section of the beam. Don't forget your internal forces—M, N, and V.

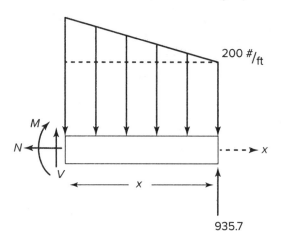

STEP 6: Write the equation for V by writing the ΣF_y equation. (Make the origin the right edge of the part.)

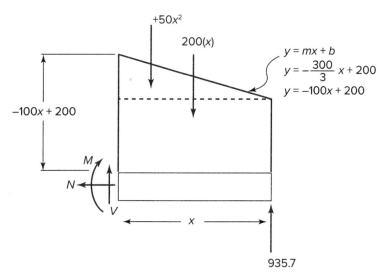

Area of triangle: $\dfrac{1}{2}(x)((-100x + 200) - 200) = 50x^2$

$\Sigma F_y = 0 = V - (50x^2) - 200(-x) + 935.7$

$V = -200x + 50x^2 - 935.7 = 50x^2 - 200x - 935.7$

Set $V = 0$ and solve for x.

$0 = 50x^2 - 200x - 935.7$

Use the quadratic equation solver on your calculator.

$x = -2.7659$

Then integrate the V equation to get the moment equation.

$$\int V = \int 50x^2 - 200x - 935.7$$

$$M = \frac{50}{3}x^3 - \frac{200x^2}{2} - 935.7x$$

Finally, substituting -2.7659 in for x, we get a maximum value of M.

$$M = 16.67(-2.7659)^3 - 100(-2.7659)^2 - 935.7(-2.7659)$$

$$M = 1,470.3$$

Now we can complete the M diagram.

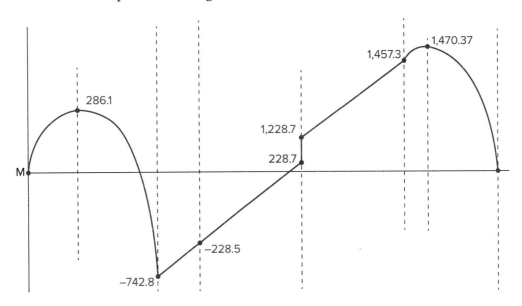

Press pause on video lesson 30 once you get to the workout problem. Only press play if you get stuck.

Draw the shear and moment diagram for a beam loaded as follows.

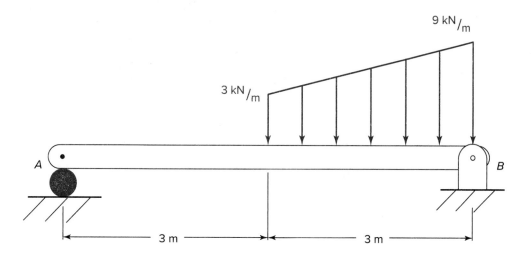

TEST YOURSELF 6.3

SOLUTION
TO TEST
YOURSELF:
Bending

For the beam shown below, sketch the shear and moment diagram. Be sure to indicate critical values on your sketches.

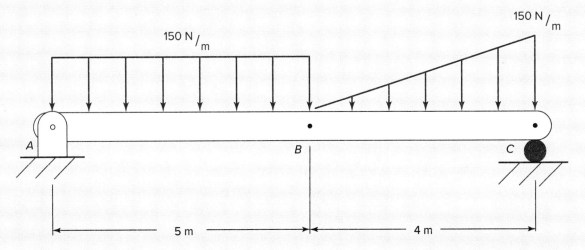

ANSWERS
Left reaction 586.1 N
Right reaction 463.9 N
$M_{max} = 1,145.8$ N·m

 **MECHANICS OF MATERIALS LESSON 31**
The Flexure Formula, Beam Bending Example

The Flexure Formula

Normal Stress (σ) caused by bending is given by the follow equations, called the Flexure Formula:

$$\sigma = \frac{-My}{I}$$
$\quad\sigma$ = Normal Stress (from fibers in tension [positive +]or compression [negative −])

M = Bending Moment (internal force in the beam)

Note: For the sign convention for the moment, positive makes you smile and negative makes you frown.

y = Distance from neutral axis to the point of interest

Note: y upward is positive and negative is downward.

I = Second area moment of inertia

$$\sigma = \frac{Mc}{I}$$
$\quad c$ = Worst-case distance (i.e., point furthest away from the neutral axis)

M and I are defined the same

For all beams experiencing bending, the outer fibers will be in tension or compression depending on the direction of bending, with the neutral axis, or centroidal axis, experiencing neither. There is no bending stress at the neutral axis (it's *zero*).

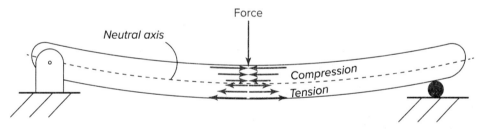

Note: The Flexure Formula technically is only valid for slender beams—a more detailed theory is the theory of elasticity, which is covered in more advanced courses.

PRO TIP

The neutral axis always passes through the centroid of the cross section for everything except for curved beams (they behave differently).

BENDING

EXAMPLE: BEAM BENDING

A beam has a cross section consisting of a semicircle welded to a T-beam. The cross-section dimensions are shown in the diagram. The moment diagram is also provided. Determine the largest compressive stress in the beam.

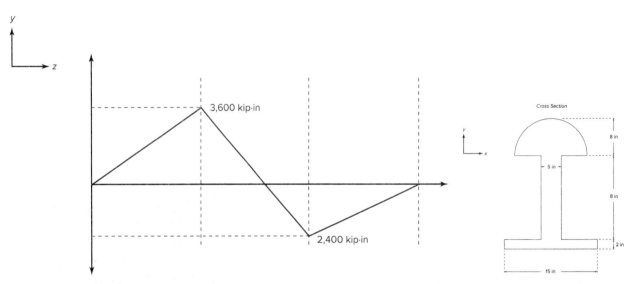

Note: You really need to spend some time and double-check your work on each step. If you get any step incorrect, it will cascade to other parts of the problem.

STEP 1:   Find the centroid of the cross section. Start by defining a datum, and identify what version of the composite equation is relevant to the problem. Use an area equation since the shape is an area. Divide the composite shape into components: two rectangles and a semicircle.

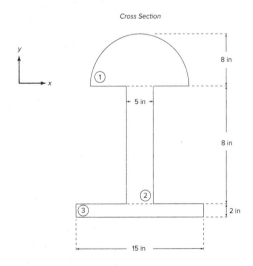

STEP 2: Fill in the following table. Recall that y_c for a semicircle is $\dfrac{4R}{3\pi}$.

Area	y_i	A_i	y_iA_i
1	$22 + (\dfrac{4(8)}{3\pi}) = 25.39$ in	$(\dfrac{1}{2})\pi 8^2 = 100.53$ in²	2,552.5 in³
2	12 in	$(5)(20) = 100$ in²	1,200 in³
3	1 in	$(2)(15) = 30$ in²	30 in³
		$\Sigma A = 230.53$ in²	$\Sigma y_iA_i = 3,782.5$ in³

$$\bar{y} = \frac{3,782.5}{230.53} = 16.4 \text{ inches upward from the bottom}$$

STEP 3: Find the I_x of the cross section.

$$I_x = I_{x_1} + I_{x_2} + I_{x_3}$$

$$I_x = (\pi\frac{8^4}{8}) - (8\frac{8^4}{9\pi}) + (25.39 - 16.4)^2(100.48)$$

$$+\frac{1}{12}((5)(20^3)) + (12 - 16.4)^2(100)$$

$$+\frac{1}{12}((15)(2^3)) + (16.4 - 1)^2(30)$$

$$I_x = 8,570.4 + 5,269.3 + 7,124.8$$

$$I_x = 20,965 \text{ in}^4$$

STEP 4: Utilize the equation $\sigma = \dfrac{-My}{I}$ to solve for the maximum compressive stress. Since in this problem, the moment along the beam changes from positive to negative, we will need to check two cases to find the maximum compressive stress:

A positive moment at the very top of the beam

$$\sigma = \frac{-My}{I} = -(3,600 \text{ kip·in})\frac{(30 \text{ in} - 16.4)}{20,965 \text{ in}^4} = -2.34 \text{ ksi}$$

A negative moment at the bottom of the beam

$$\sigma = \frac{-My}{I} = -(-2,400 \text{ kip·in})\frac{-16.4}{20,965 \text{ in}^4} = -1.88 \text{ ksi}$$

Worse-case compressive stress is 2.34 ksi

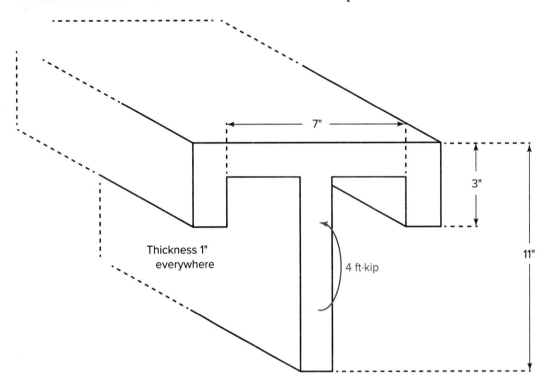 Press pause on video lesson 31 once you get to the workout problem. Only press play if you get stuck.

Find the maximum tensile stress and maximum compressive stress due to beam bending.

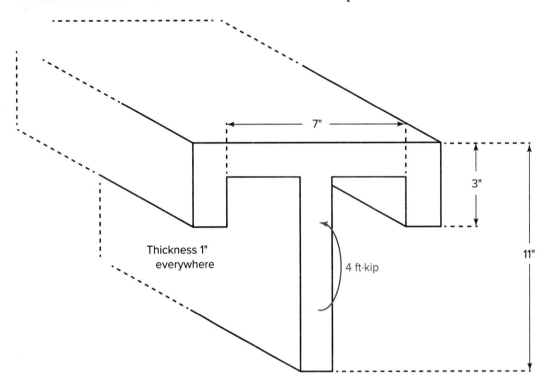

7"

3"

11"

Thickness 1"
everywhere

4 ft·kip

SOLUTION TO TEST YOURSELF: Bending

✓ TEST YOURSELF 6.4

Two planks are firmly attached to form the cross section shown below. If a positive moment of 4.5 kN•m is applied, what are the maximum tensile and compressive stresses, and where do they occur?

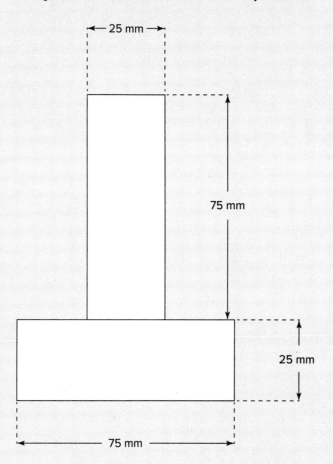

25 mm

75 mm

25 mm

75 mm

MECHANICS OF MATERIALS LESSON 32
Never Get the Polar and Area Moment of Inertia Backward Again!

Reminder: J and I look alike, but they are different. In the previous lesson, we used the "polar moment of inertia (J)" in equations, and in this level, we will mainly be using the "area moment of inertia (I)," and the equations for both are very similar and easy to get confused.

I tells you the resistance to bending;

J tells you the resistance to torsion.

With the trick in the pitfall below, you will never make that mistake again!

BENDING

PITFALL

Beware: It is very easy to get the equations for J and I mixed up. Here's a trick to remember—the line that forms the J has two ends, and I has four ends.

$$J = \frac{\pi}{2}(r)^4 \qquad I = \frac{\pi}{4}(r)^4$$

MECHANICS OF MATERIALS LESSON 33
The Flexure Formula with Shear Moment Diagram

Press pause on video lesson 33 once you get to the workout problem. Only press play if you get stuck.

Draw the shear and moment diagram for the loaded beam below. Find the absolute largest normal bending stress in the beam.

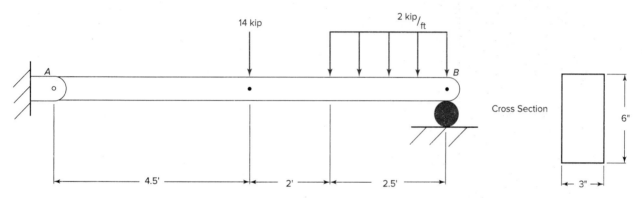

14 kip

2 kip/ft

A

B

Cross Section

6"

4.5'   2'   2.5'

3"

TEST YOURSELF 6.5

SOLUTION TO TEST YOURSELF: Bending

A beam is loaded as shown below. The cross section is also shown. Compute the maximum tensile and compressive bending stress in the cross section due to this loading.

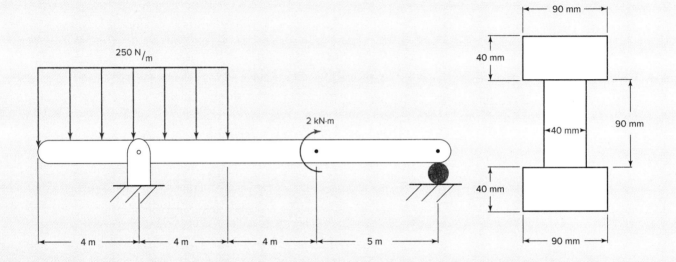

ANSWER
Maximum bending stress 5.03 MPa

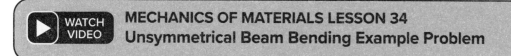

Unsymmetric Bending

When you hear "unsymmetric bending," think bending around an axis other than the x, y, or z axis.

The easiest way to approach these problems is to simply resolve forces and moments into x, y, and z components.

Note: Don't forget that you can resolve a moment into components the same way you can resolve forces into components.

Once moments and forces on bodies are resolved into components, symmetric bending forces/moments become easier to handle.

Finally, the best approach to solve these problems is to draw an "effect map" of the cross section of the beam containing the point of interest.

"Effect Map"

To construct these "effect maps," first draw the cross section of the beam containing the point of interest.

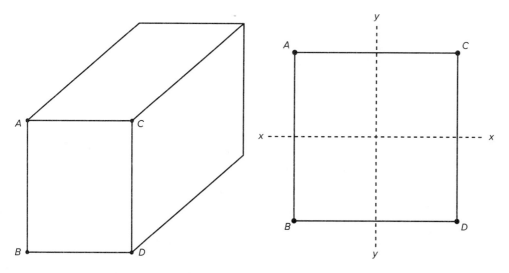

(Remember for bending, at least for now, we are only talking about normal stresses created $[\sigma]$.)
Add the "effect" from forces/moments to your map. The "effects" are either normal forces, causing $\sigma = \dfrac{P}{A}$ type stress, or bending moments, causing $\sigma = \dfrac{Mc}{I}$ type stress.

- Draw normal forces as a big dot in the center and denote whether it is $(+)$ tension or $(-)$ compression.
- Draw bending moments across the face in the direction of the bending.

Example of what an effect map might look like:

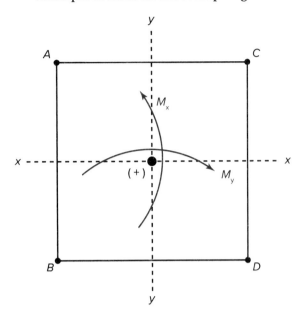

Using your "effect map," calculate the total stress at each point due to compression/tension forces and then bending moments.

PRO TIP

Here's where your small pieces of "pool noodle" really can come in handy to see (visualize) whether a point is on the compression or tension side of bending on the beam, paying close attention to signs.

$$\sigma_{total} = \frac{P}{A} + \frac{Mc}{I}$$ Recall tension (+) and compression (−).

BENDING

EXAMPLE: UNSYMMETRICAL BEAM BENDING

Find the bending stress at corners *A* and *B*.

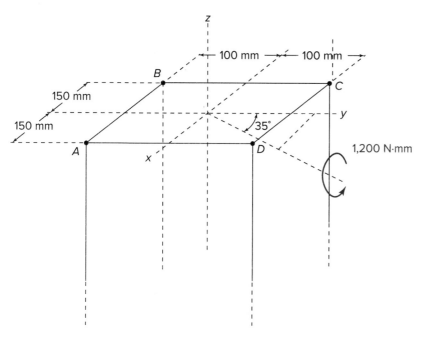

STEP 1: Resolve the moment into components. This is done using the same method you used to resolve forces into components.

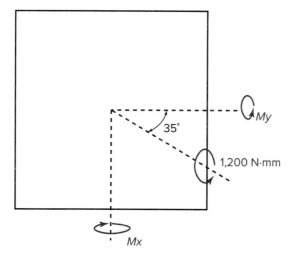

$$M_y = 1{,}200\ cos(35°) = 982.98\ \text{N•mm}$$
$$M_x = 1{,}200\ sin(35°) = 688.3\ \text{N•mm}$$

STEP 2: Add bending moment components to "effect map."

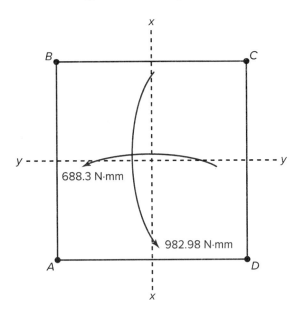

STEP 3: Now that we know the effects from the moments, we can analyze points A, B, C, and D. But first, we must calculate the moment of inertia about the x-axis and the y-axis.

$$I_{xx} = \frac{1}{12}(300)(200)^3 = 200{,}000{,}000 \text{ mm}^4$$

$$I_{yy} = \frac{1}{12}(200)(300)^3 = 450{,}000{,}000 \text{ mm}^4$$

STEP 4: Add the components of stress together for each point.

$$\sigma_A = -\frac{M_y(x)}{I_{yy}} - \frac{M_x(y)}{I_{xx}}$$

$$\sigma_A = -\frac{(982.98 \text{N·mm})(150 \text{ mm})}{450{,}000{,}000 \text{ mm}^4} - \frac{(688.3 \text{N·mm})(100 \text{ mm})}{200{,}000{,}000 \text{ mm}^4}$$

$$\sigma_A = -327.66 \text{ Pa} - 344.15 \text{ Pa} = -671.81 \text{ Pa (compressive stress)}$$

$$\sigma_B = \frac{M_y(x)}{I_{yy}} - \frac{M_x(y)}{I_{xx}}$$

$$\sigma_B = \frac{(982.98 \text{N·mm})(150 \text{ mm})}{450{,}000{,}000 \text{ mm}^4} - \frac{(688.3 \text{N·mm})(100 \text{ mm})}{200{,}000{,}000 \text{ mm}^4}$$

$$\sigma_B = 327.66 \text{ Pa} - 344.15 \text{ Pa} = -16.49 \text{ Pa (compressive stress)}$$

BENDING

Press pause on video lesson 34 once you get to the workout problem. Only press play if you get stuck.

Find the normal stresses at points *A* and *B*. Note 850 N·m is on the *x-y* plane.

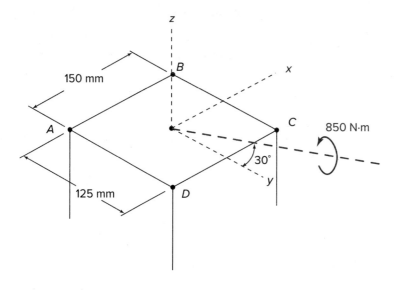

✓ TEST YOURSELF 6.6

SOLUTION TO TEST YOURSELF: Bending

The 4 m long cantilevered beam shown below has a square cross section 100 mm on each side. A 1,500 kN point load is applied to the centroid of the cross section in the xy-plane at an angle of 60°. In addition, an axial load of 100 kN is applied at the end of the beam as shown. Compute the maximum normal stress in the four corners of the cross section. Ignore transverse shear. You will learn about principal stresses later in the book.

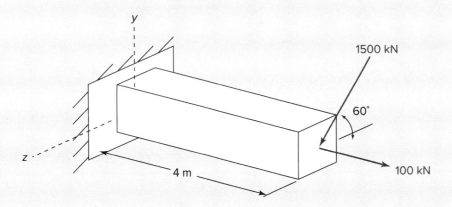

ANSWERS
Tension (max) 59.2 MPa
Compression (max) 39.2 MPa

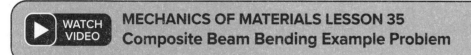

Composite Beam Bending

Occasionally, beams are constructed from more than one material. A very common example of this is a steel beam with wood attached.

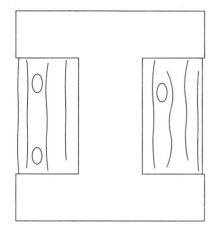

Even though beams may be constructed of two different materials, the Flexure Formula ($\sigma = \dfrac{Mc}{I}$) is only valid for one material at a time. So, in order to use the Flexure Formula with these composite beams, we must convert one material to the other.

The steel on these beams is carrying the bulk of the load, and the wood allows construction attachment using nails or screws.

Note: It does not matter which material is "converted" to the other material.

Material Conversion Factor

To convert one material to another, we compare the modulus of elasticity of each material.

$$n = \frac{E_1}{E_2}$$

We use this factor to change the width by the amount of the conversion factor so that we can pretend it is all the same material for stress calculations. Then at the end, we take the stress calculated and convert it so that it is the stress in the actual material.

BENDING

PRO TIP

Pick the material with the higher "*E*" to be the numerator on the conversion factor so that *n* is always greater than 1. Then when using the material conversion factor, don't change the height as that messes with your "*c*" distance in the bending stress equation. You can only change the width of the smaller stiffness material!

PITFALL

Don't forget to reconvert the transformed material. You must take the stress calculated for the transformed material and multiply it by *n*. There is no need to convert the base material (the one you didn't convert).

BENDING

EXAMPLE: COMPOSITE BEAM BENDING

Find the maximum normal bending stress in the steel and in the red brass in the composite beam shown below.

$E_{ST} = 200$ GPa

$E_{RB} = 101$ GPa

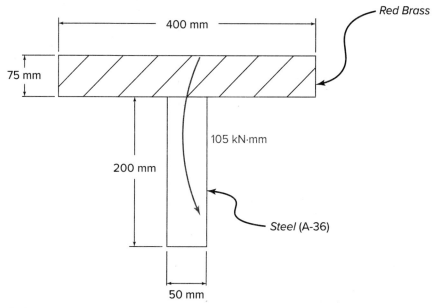

STEP 1: Calculate n (conversion factor) by dividing the stronger material by the weaker. This will always give us a value bigger than 1.

$$n = \frac{200}{101} = 1.98$$

STEP 2: Convert the weaker material to the stronger material and redraw. Remember we can make this part as wide as needed, but you cannot make the part any taller. (We cannot vary the height from the original.)

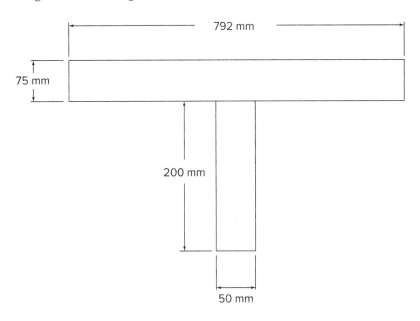

STEP 3: Calculate the new centroid ($\bar{y}$).

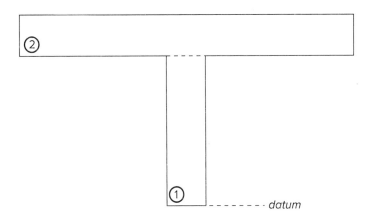

Area	y_i	A_i	y_iA_i
1	100	10,000	1,000,000
2	237.5	59,400	14,107,500
		ΣA = 69,400 mm²	Σy_iA_i = 15,107,500 mm³

$$\bar{y} = \frac{15,107,500}{69,400} = 217.7 \text{ mm}$$

STEP 4: Calculate area moment of inertia (I).

$$I_{xx} = \frac{1}{12}bh^3 + Ad^2$$

$$= [\frac{1}{12}(50)(200)^3 + (10,000)(117.7)^2] + [\frac{1}{12}(792)(75)^3 + (59,400)(19.8)^2]$$

$$= 171,866,233.3 + 51,130,926$$

$$= 222,997,159.3 \text{ mm}^4$$

STEP 5: Use the Flexure Formula to calculate the normal stress in the beam due to bending. Let's find the stress in the steel first. Since the steel is on the bottom of the part, the stress will be negative since it is in compression (according to the applied moment).

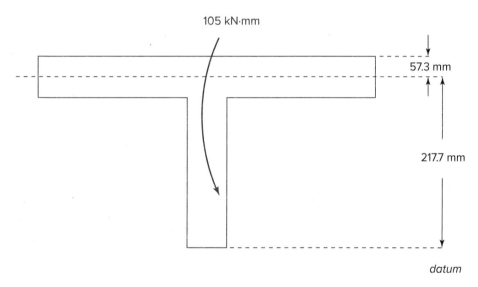

$$\sigma_{ST} = \frac{-My}{I} = \frac{-(-105,000 \text{ N·mm})(-217.7 \text{ mm})}{222,997,159.3 \text{ mm}^4} = 0.1025 \frac{\text{N}}{\text{mm}^2}$$

$$\sigma_{ST} = -102.5 \text{ kPa (compression)}$$

STEP 6: Calculate stress for the red brass (top of the part).

$$\sigma_{BR} = \frac{-(-105,000 \text{ N·mm})(57.3 \text{ mm})}{222,997,159.3 \text{ mm}^4} = 0.02698 \frac{\text{N}}{\text{mm}^2}$$

$$\sigma_{BR} = 26.98 \text{ kPa}$$

(This value is for steel. Remember, we converted the brass to steel, now we have to convert back to red brass by multiplying it by our "n" value.)

$$\sigma_{BR} = 26.98 \text{ kPa} \times 1.98 = 53.4 \text{ kPa}$$

Press pause on video lesson 35 once you get to the workout problem. Only press play if you get stuck.

For the loading scenario below: (a) find the maximum stress in the brass and the maximum stress in the steel, and (b) find the stress in each material at the bond line.

The Young's modulus of brass is 100 GPa, and the Young's Modulus of steel is 200 GPa.

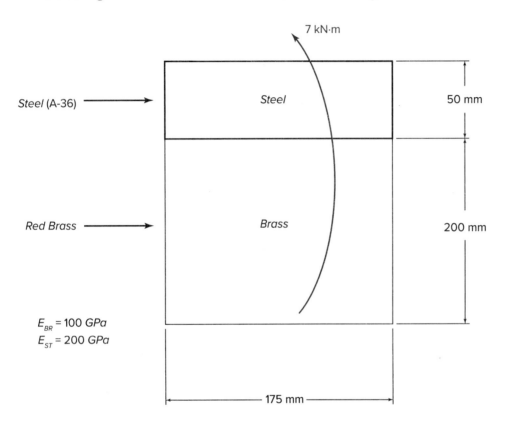

7 kN·m

Steel (A-36)

Steel

50 mm

Red Brass

Brass

200 mm

$E_{BR} = 100 \; GPa$
$E_{ST} = 200 \; GPa$

175 mm

✓ TEST YOURSELF 6.7

The 8-ft composite cross section beam carries a 100 lb load at the tip. Material *A* has an elastic modulus of 7×10^6 psi, while material *B* has an elastic modulus of 17.5×10^6 psi. What is the maximum bending stress in the beam?

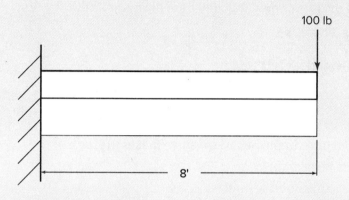

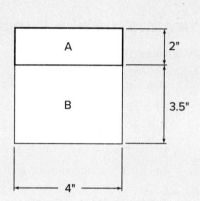

Shear and Moment Diagram Review

- There are two methods to draw shear and moment diagrams: (1) graphic method and (2) equation method.
  - It is helpful to remember the order of the lines when drawing these diagrams as the moment diagram is the integral of the shear curve.

Flexure Formula to Calculate Bending Stress

- To calculate bending stress:

$$\sigma = \frac{-My}{I}$$ σ = Normal Stress (from fibers in tension [positive +]or compression [negative −])

M = Bending Moment (internal force in the beam) (positive makes the beam "smile" and negative makes the beam "frown")

y = Distance from neutral axis to the point of interest

I = Second area moment of inertia

$$\sigma = \frac{Mc}{I}$$ c = Worst-case distance (i.e., point furthest away from the neutral axis)

M and I are defined the same

Unsymmetric Beam

- When you hear "unsymmetric bending," think bending around an axis other than the x, y, or z axis.
- The easiest way to approach these problems is to simply resolve forces and moments into x, y, and z components.
- *Note:* Don't forget that you can resolve a moment into components the same way you can resolve forces into components.
- The best approach to solve these problems is to draw an "effect map" of the cross section of the beam containing the point of interest.

Composite Beam

- Composite beams are when beams are constructed from more than one material.
- Even though beams may be constructed of two different materials, the Flexure Formula $(\sigma = \dfrac{Mc}{I})$ is only valid for one material at a time. So, in order to use the Flexure Formula with these composite beams, we must convert one material to the other through the material conversion factor.

$$n = \frac{E_1}{E_2}$$ E_1 = Young's Modulus of Material 1 (stiffer material)

E_2 = Young's Modulus of Material 2 (weaker material)

Equations Learned in this Level

Flexure Formula

$$\sigma = \frac{-My}{I}$$

$$\sigma = \frac{Mc}{I}$$

Material Conversion Factor (for Composite Beams)

$$n = \frac{E_1}{E_2}$$

PRO TIPS

Flexure Formula

- The neutral axis always passes through the centroid of the cross section for everything except for curved beams (they behave differently).
- Here's where your small pieces of "pool noodle" really can come in handy to see (visualize) whether a point is on the compression or tension side of bending on the beam, paying close attention to signs.

$$\sigma_{total} = \frac{P}{A} + \frac{Mc}{I}$$ Recall tension (+) and compression (−).

- Pick the material with the higher "E" to be the numerator on the conversion factor so that n is always greater than 1. Then when using the material conversion factor, don't change the height as that messes with your "c" distance in the bending stress equation. You can only change the width of the smaller stiffness material!

PITFALLS

Flexure Formula

- *Beware:* It is very easy to get the equations for J and I mixed up. Here's a trick to remember—the line that forms the J has two ends, and I has four ends.

$$J = \frac{\pi}{2}(r)^4 \qquad I = \frac{\pi}{4}(r)^4$$

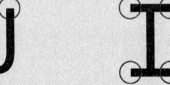

- Don't forget to reconvert the transformed material. You must take the stress calculated for the transformed material and multiply it by n. There is no need to convert the base material (the one you didn't convert).

Level 7
Transverse Shear

Transverse Shear

Shear Flow

WATCH VIDEO **MECHANICS OF MATERIALS LESSON 36**
What Is Transverse Shear? Explained

What Is Transverse Shear?

At this point we are familiar with the concept of shear, and we think of it as a "tearing" of material due to two opposing, equal but opposite, forces as shown below.

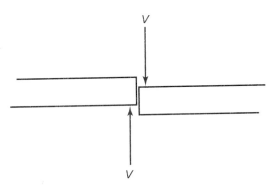

This was true when the shear force, V, was present causing shear stress, $\tau = \dfrac{V}{A}$.

However, shear can also be caused by beam bending. As a beam experiences bending, the fibers in the beam slide one layer relative to the next as shown in the illustration below.

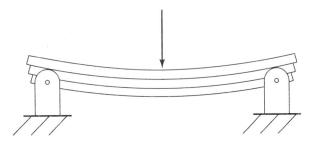

If you can imagine a beam as being built of layers of material, you can visualize how each layer slips relative to the layer above or below it. This "slipping" is called transverse shear.

The word *transverse* means across or crosswise. In other words, we think a downward force would cause a vertical shear, but it also causes a horizontal (transverse) shear. This seems counterintuitive or maybe like a completely new stress, but it really isn't. Think of yourself as a small piece of material on the edge of a shear plane. This small piece of material is called a 3D stress element.

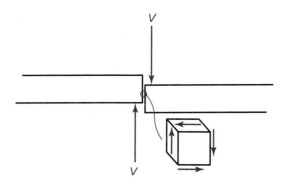

That 3D stress element is going to have shear stresses on its faces

If we realize that these stress elements are not spinning, we can easily visualize that if there is a vertical shear stress, there would have to be a horizontal shear stress as well, in order to be in static equilibrium.

Another difficult to understand concept is that transverse shear magnitudes change at different points in the beam.

Again, imagine yourself as a stress element. The farther into the beam you go, approaching the neutral axis, the more material you would have above you. The more material you have above you, the more "opposing" material you have to bending stress. With this in mind, it is easy to see that at the outside fibers of the beam, there would be nothing above you and, therefore, no one to oppose the shear stress, so it would be equal to zero.

The transverse shear stress is typically maximum at the neutral axis (N.A.) as that is where you would have maximum material above you. Therefore, the magnitude of the shear stress on a rectangular beam would look like the following:

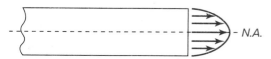

Equation for Transverse Shear

The equation for transverse shear is $\tau = \dfrac{VQ}{It}$.

V = shear force (from the shear diagram)

Q = first area moment of inertia (discussion to follow in Lesson 37)

I = second area moment of inertia

t = total thickness at the point of interest

So why are $\tau = \dfrac{V}{A}$ and $\tau = \dfrac{VQ}{It}$ not the same value?

$\tau = \dfrac{V}{A}$ is *average* shear stress where $\tau = \dfrac{VQ}{It}$ is shear stress due to bending at a specific point.

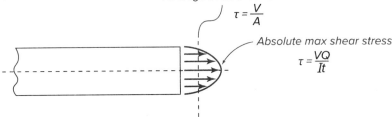

Average of all shear stresses

$\tau = \dfrac{V}{A}$

Absolute max shear stress

$\tau = \dfrac{VQ}{It}$

As engineers, we are interested in shear stress at particular points, such as where two different geometries intersect or the point in the beam where the absolute shear stress is maximum. You could also be interested in knowing the shear stress when you are planning on mating two different materials, and you need to select the correct glue, weld material, nails, etc. that can survive this stress.

CHALLENGE QUESTION

When do you use $\tau = \dfrac{V}{A}$ versus $\tau = \dfrac{VQ}{It}$?

Where does maximum shear stress occur?

ANSWER

$\tau = \dfrac{V}{A}$ is the average shear stress at the plane of interest. It doesn't take into account the distribution of the shear stress, whereas $\tau = \dfrac{VQ}{It}$ is the stress at any particular point of interest within the beam.

The maximum transverse shear stress typically occurs at the neutral axis.

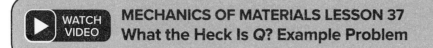

MECHANICS OF MATERIALS LESSON 37
What the Heck Is Q? Example Problem

What the Heck Is *Q*?

In order to use the transverse shear equation, we will need to calculate the first area moment of inertia, known as *Q*. Calculating *Q* is, perhaps, the most confusing concept in Mechanics of Materials.

The equation for *Q* is: $Q = \Sigma y_i A_i$

This may look familiar as it is the top half of the centroid equation.

- If the point of interest is above the neutral axis, it is best to use the area *above* the point of interest.
- If the point of interest is below the neutral axis, it is best to use the area *below* the point of interest.

Once you have identified the area of interest either above or below the point of interest, then to calculate *Q* typically we break this into several common shapes whose centroid is known or can be looked up in a table.

A_i: A component of the composite shape above or below the neutral axis.

y_i: Simply the $\bar{y}$ centroid of the "shaded region" to the neutral axis of the entire cross section.

The reason for the Σ (summation) is that sometimes these areas are composite shapes. See the example problem that follows.

EXAMPLE: Q CALCULATIONS

Calculate Q at points A, B, and C.

Note: The max Q exists at the neutral axis. When calculating Q at this point (point A), you can choose the area above the neutral axis or the area below. You will arrive at the same value. Let's calculate Q both ways to prove this point.

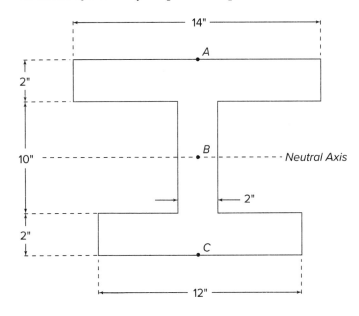

STEP 1: Calculate $\bar{y}$ in order to find the location of the neutral axis for the entire cross section.

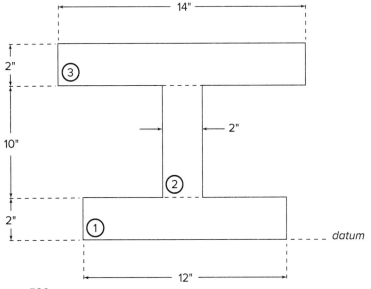

Area	y_i	A_i	y_iA_i
1	1 in	24 in²	24 in³
2	7 in	20 in²	140 in³
3	13 in	28 in²	364 in³
		ΣA_i = 72 in²	Σy_iA_i = 528 in³

$$\bar{y} = \frac{528}{72} = 7.33"\text{ upward from bottom}$$

STEP 2: Calculate Q_A (Q at the neutral axis).

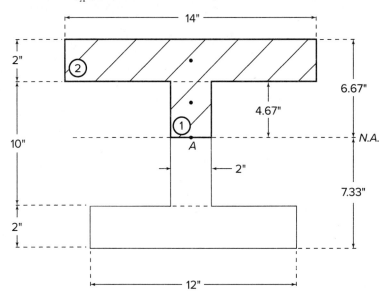

Let's calculate Q_A above the neutral axis and below the neutral axis to prove they are the same.

$$Q_A = [(\frac{4.667}{2})(4.667)(2) + (5.667)(2)(14)]$$

$$Q_A = 180.44 \text{ in}^3$$

Note: Even though the units here are inches cubed, this does *not* indicate a volume! It is simply area (in²) times length (in) giving in³.

Now for the area below the neutral axis:

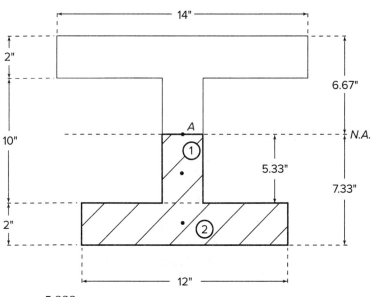

$$Q_A = [(\frac{5.333}{2})(5.333)(2) + (6.333)(12)(2)]$$

$$Q_A = 180.44 \text{ in}^3$$

STEP 3: Calculate Q_B. Remember, since point B is below the neutral axis, the area we use here is *below* point B (the point of interest).

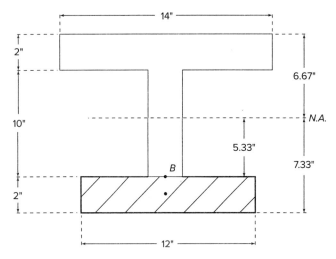

$$Q_B = [(5.33 + 1)(2)(12)] = 151.92 \text{ in}^3$$

STEP 4: Calculate Q_C. Since point C is at the outermost fibers of the beam, there is no area *above* that point. Therefore, $Q_C = 0$.

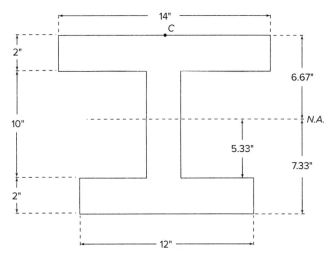

Understand that if $Q = 0$, then transverse shear stress $(\tau = \dfrac{VQ}{It})$ at C is also equal to zero.

PRO TIP

Just like in previous levels, using the transverse shear equation successfully depends on your units all being compatible!
Make sure to use base units (N, mm for metric, in, kips for Imperial), so that everything cancels out.
Avoid this biggest mistake!

Press pause on video lesson 37 once you get to the workout problem. Only press play if you get stuck.

Find the Q at points A, B, C, and D. Assume point C is in the center of the upper flange.

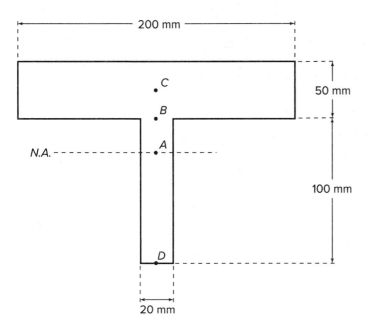

TEST YOURSELF 7.1

SOLUTION TO TEST YOURSELF:
Transverse Shear

Compute the *maximum* first moment of the area (Q) for the shape shown.

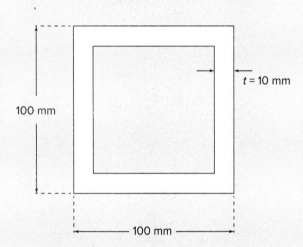

$t = 10$ mm

100 mm

100 mm

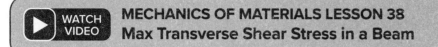

▶ WATCH VIDEO | **MECHANICS OF MATERIALS LESSON 38**
Max Transverse Shear Stress in a Beam

To find the maximum shear stress, you first have to find where the shear force (*V*) in the beam is maximum. This will require you to draw the shear diagram. Then, once you find the maximum *V* you will need to find the location of the point on that plane where the bending shear stress is maximum. Most of the time, this will occur at the neutral axis (see the beam-cracked image below) unless the beam is "fatter" as you approach the neutral axis such that "*t*" (in the bending shear stress equation) is becoming bigger. Then you might have to check a few places. (For example, in the beam in the diagram below, you would want to check the shear stress at *A*, *B*, and *C*.)

Notice the beam is cracked along the neutral axis in the transverse direction.

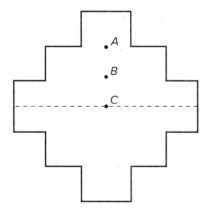

Beam where the max shear stress might not occur at the neutral axis.

TRANSVERSE SHEAR

EXAMPLE: MAX TRANSVERSE SHEAR STRESS

For the following loaded beam, find absolute maximum shear stress.

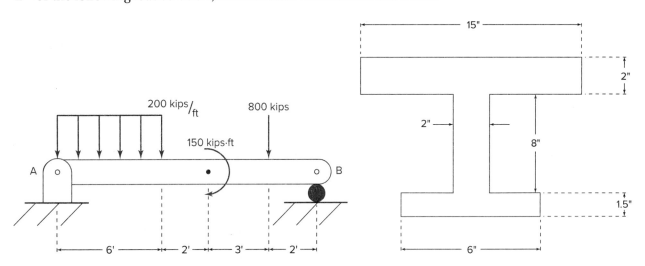

STEP 1: Find maximum shear force, V, on this loaded beam. Find global equilibrium and plot the V diagram.

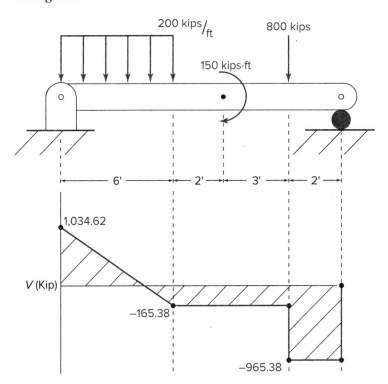

Global Equilibrium:

$\Sigma M_A = 0 = -1{,}200(3) - 150 - 800(11) + B_y(13)$

$B_y = 965.38 \text{ kips}$

$\Sigma F_y = 0 = 965.38 - 800 - 1{,}200 + A_y$

$A_y = 1{,}034.62 \text{ kips}$

Therefore, maximum shear force on this beam is $V = 1{,}034.62$ kips.

STEP 2: Calculate $\bar{y}$ for the cross section of the beam in order to find the location of the neutral axis.

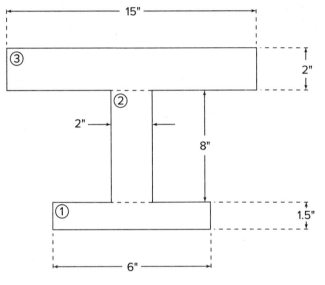

Area	y_i	A_i	y_iA_i
1	0.75 in	9 in²	6.75 in³
2	5.5 in	16 in²	88.0 in³
3	10.5 in	30 in²	315.0 in³
		$\Sigma A_i = 55$ in²	$\Sigma y_iA_i = 409.75$ in³

$$\bar{y} = \frac{409.75}{55} = 7.45\text{" upward from bottom}$$

STEP 3: Calculate the second area moment of inertia (I). Our shape is a composite shape containing rectangles, and we need to use the parallel axis theorem.

$$I = \frac{1}{12}(6)(1.5)^3 + (6)(1.5)(7.45 - 0.75)^2$$

$$+ \frac{1}{12}(2)(8)^3 + (2)(8)(7.45 - 5.5)^2$$

$$+ \frac{1}{12}(15)(2)^3 + (15)(2)(7.45 - 10.5)^2$$

$I = [1.6875 + 404.01] + [85.333 + 60.84] + [10.00 + 279.075]$

$I = 840.95 \text{ in}^4$

STEP 4: Calculate the first area moment of inertia (Q). Since we are looking for maximum shear stress, calculate Q about the neutral axis.

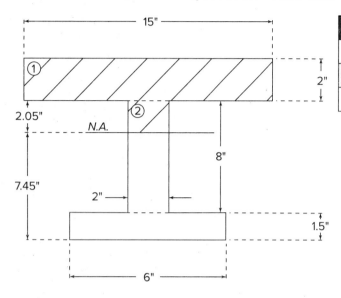

Area	y_i	A_i	y_iA_i
1	(1 + 2.05) in	30 in²	91.5 in³
2	(2.05/2) in	4.1 in²	4.2025 in³
			Q = 95.7025 in³

STEP 5: Determine thickness (t).

In this case, $t = 2"$ (the thickness at the neutral axis).

STEP 6: $\tau = \dfrac{VQ}{It}$

$$= \frac{(1{,}034{,}620 \text{ lb})(95.7025) \text{ in}^3}{(840.95) \text{ in}^4 \, (2) \text{ in}}$$

$$\tau = 58{,}870 \text{ psi}$$

▶ Press pause on video lesson 38 once you get to the workout problem. Only press play if you get stuck.

Plot the transverse shear across the face of the I-beam.

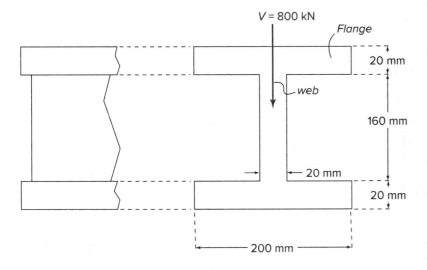

TEST YOURSELF 7.2

SOLUTION TO TEST YOURSELF: Transverse Shear

1. For the beam loaded as shown, compute the maximum bending stress and shear stress.

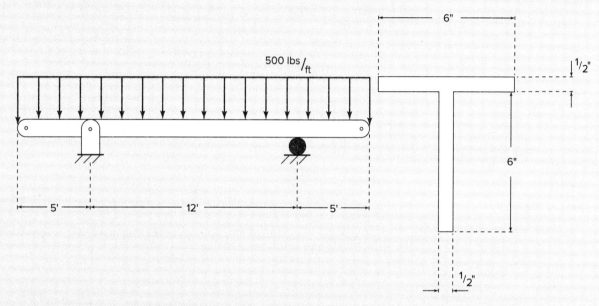

2. A beam cross section is composed of six 1 inch thick blocks glued together as shown. Which interface will be the weakest?

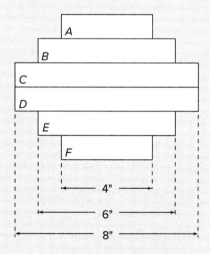

TRANSVERSE SHEAR

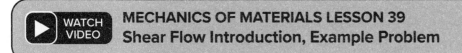

MECHANICS OF MATERIALS LESSON 39
Shear Flow Introduction, Example Problem

What Is Shear Flow?

Shear flow is a concept that pairs perfectly with transverse shear, in that it is the resistance to transverse shear. Imagine our transverse shear planes that want to slide one relative to the other.

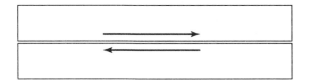

Shear flow would be the force per length that would be required to prevent that shearing from occurring. The equation for shear force "q" is nicknamed the "Vicky Equation":

$q = \dfrac{VQ}{I}$ q = shear flow in florce/length (i.e., $\dfrac{N}{m}, \dfrac{kip}{in}$)

V = shear force (obtained from the shear diagram)

Q = first area moment of inertia ($\Sigma y_i A_i$)

I = second area moment of inertia

We use shear flow for instances of determining the number of fasteners per length (nails, screws) or the strength of glue needed to prevent slipping on shear planes.

Another way to write the equation for shear force is:

$q = \dfrac{f}{s}$ f = force fastener can resist in shear

s = spacing of fasteners (nails or screws)

Definitely, the trickiest thing about these calculations is coming up with Q. The best way to think about these problems, especially coming up with Q, is to think about the joint that is screwed, glued, or nailed and how you could make that joint fail. In other words, which piece of the assembly, or built up beam, would I have to "tear off" to make my fasteners fail?

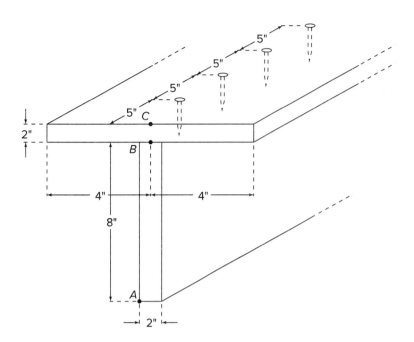

In this example, to make the nails fail at joint *B*, I would have to "tear off" that top board. Therefore, "*Q*" would be that top board's area, above point *B*. As compound beams become more complex, *Q* is harder to imagine. Just remember this: the shear planes (places where you want the beam to fail) must be symmetrical. If I have a vertical shear plane on one side of the beam, so it has to have a vertical shear plane on the other side of the beam (further discussed in lesson 40).

EXAMPLE: SHEAR FLOW

If each nail in the structure below can resist a shear force of 200 lbs, find the necessary spacing of the nails.

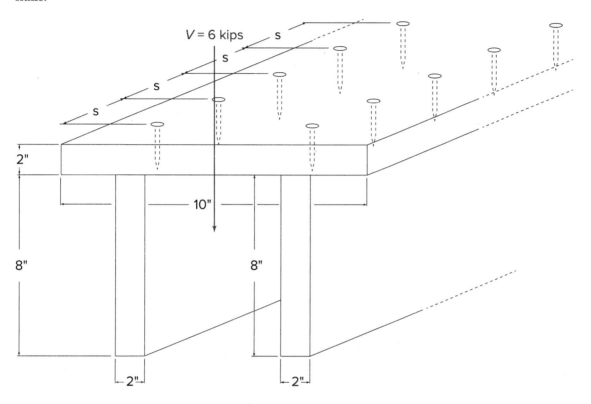

STEP 1: Calculate the centroid of the shape in order to find the neutral axis.

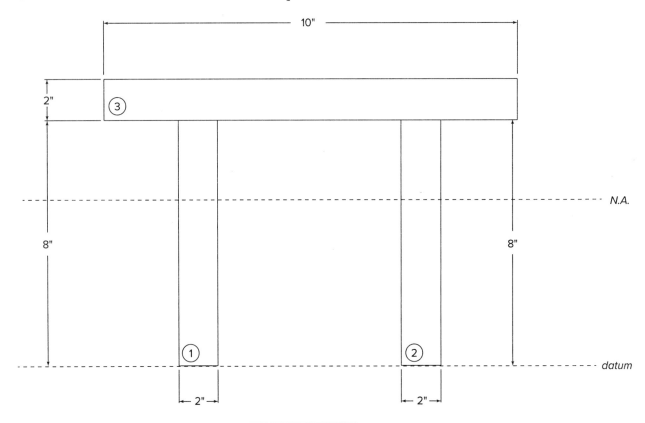

Area	y_i	A_i	yA_i
1	4 in	16 in²	64 in³
2	4 in	16 in²	64 in³
3	9 in	20 in²	180 in³
		$\Sigma A_i = 52$ in²	$\Sigma y_i A_i = 308$ in³

$$\bar{y} = \frac{308}{52} = 5.923"$$

STEP 2: Calculate the second area moment of inertia (I).

$$I = [\frac{1}{12}(2)(8)^3 + (16)(1.923)^2]\,(2)$$

$$[\frac{1}{12}(10)(2)^3 + (20)(3.077)^2]$$

$$I = 485.03 \text{in}^4$$

STEP 3: Calculate Q. Again, think about the part of the beam we would have to "tear" away to make the nails fail . . . in this case, the top board.

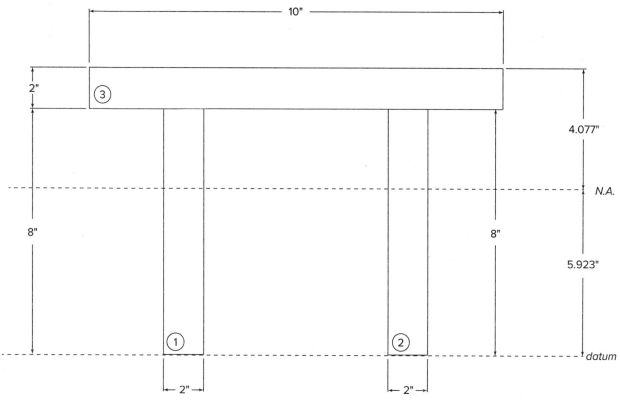

$$Q = \Sigma y_i A_i$$
$$= (3.077)(2)(10)$$
$$= 61.54 \text{ in}^3$$

STEP 4: Calculate shear flow "q".

$$q = \frac{VQ}{I} = \frac{(6{,}000 \text{ lbs})(61.54 \text{ in}^3)}{485.03 \text{ in}^4} = 761.3 \text{ lbs/in}$$

STEP 5: Finally, calculate nail spacing.

$$q = \frac{f}{s}$$

For f, remember we have two nails per row that would have to fail, so our equation will be:

$$q = \frac{2f}{s} \text{ or } s = \frac{2f}{q}$$

$$s = \frac{2(200 \text{ lbs})}{761.3 \text{ lbs/in}} = 0.525 \text{ in}$$

$\therefore$ Our minimum nail spacing has to be 0.525 in.

Press pause on video lesson 39 once you get to the workout problem. Only press play if you get stuck.

If the maximum shear force a nail can withstand is 15 kN, find the spacing of the nails along the beam.

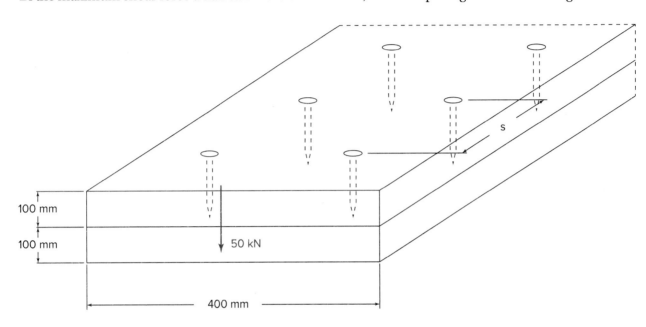

100 mm

100 mm

50 kN

400 mm

SOLUTION
TO TEST
YOURSELF:
Transverse
Shear

✓ TEST YOURSELF 7.3

The 8 ft built-up cantilever beam, with the cross section shown, is loaded at the end with a 250 lb force as shown. If the fasteners can hold 200 lb each, what spacing should be used?

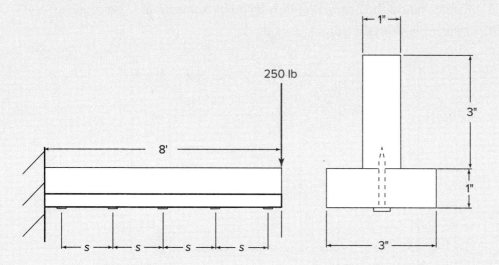

MECHANICS OF MATERIALS LESSON 40
Harder Built-Up Member Shear Flow Problem

Thus far we have only looked at the vertical shear plane with everything being symmetric. However, we haven't shown how to deal with the nails being pointed in different directions (not just up and down but also horizontal). If we have a horizontal shear plane on one side of the beam, we must have a horizontal shear plane on the other side of the beam. Similarly if you have a vertical shear plane on one side, you must have a vertical shear plane on the other side of the beam.

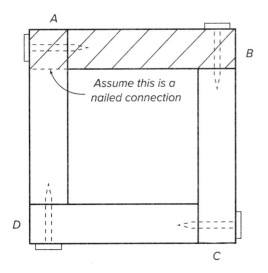

Horizontal shear plane failure at joint *B*
(*Q* is shown shaded.)

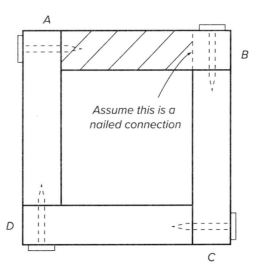

Vertical shear plane failure at joint A
(*Q* is shown shaded.)

Even though only one side is nailed together, we have to assume that the wood is at least as strong as the nail on the other side, and we would calculate Q as if there were two nails (one on each side) holding the structure together.

▶ Press pause on video lesson 40 once you get to the workout problem. Only press play if you get stuck.

Find the spacing for the nails at point *A* and at point *B*. Note that the nails can only support 50 # of shear force.

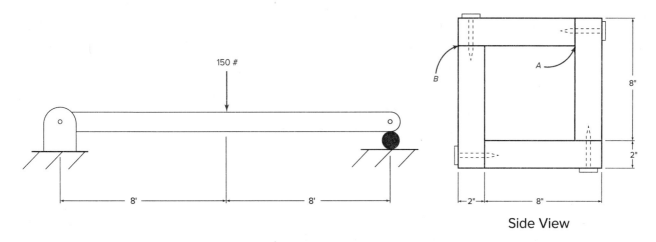

Side View

Transverse Shear

- The equation for transverse shear is $\tau = \dfrac{VQ}{It}$
  - V = shear force (from the shown diagram)
  - Q = first area moment of inertia
  - I = second area moment of inertia
  - t = total thickness at the point of interest
- The equation for Q is: $Q = \Sigma y_i A_i$
  - This may look familiar as it is the top half of the centroid equation.
    - If the point of interest is above the neutral axis, it is best to use the area *above* the point of interest.
    - If the point of interest is below the neutral axis, it is best to use the area *below* the point of interest.
  - A_i: A component of the composite shape above or below the neutral axis.
  - y_i: simply the $\bar{y}$ centroid of the "shaded region" to the neutral axis of the entire cross section.
  - The reason for the Σ (summation) is that sometimes these areas are composite shapes.

Shear Flow Introduction

- Shear flow is the force per length that would be required to prevent that shearing from occurring
- $q = \dfrac{VQ}{I}$
  - q = shear flow in florce/length (i.e., $\dfrac{N}{m}, \dfrac{kip}{in}$)
  - V = shear force (obtained from shear diagram)
  - Q = first area moment of inertia ($\Sigma y_i A_i$)
  - I = second area moment of inertia
- We use shear flow for instances of determining the number of fasteners per length (nails, screws) or the strength of glue needed to prevent slipping on shear planes.
- $q = \dfrac{f}{s}$
  - f = force fastener can resist in shear
  - s = spacing of fasteners (nails or screws)

KEY TAKEAWAYS

Equations Learned in this Level

Transverse Shear Formula

$$\tau = \frac{VQ}{It}$$

First Area Moment of Inertia

$$Q = \Sigma y_i A_i$$

Shear Flow Formulas

$$q = \frac{VQ}{I}$$

$$q = \frac{f}{s}$$

PRO TIP

Transverse Shear
- Just like in previous levels, using the transverse shear equation successfully depends on your units all being compatible!

    Make sure to use base units (N, mm for metric, in, kips for Imperial), so that everything cancels out.

    Avoid this biggest mistake!

Level 8
Combined Loading

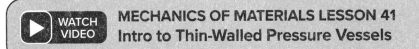

Introduction to Thin-Walled Pressure Vessels

There are two basic types of pressure vessels that we will encounter. One is a cylindrical pressure vessel, and the other is spherical.

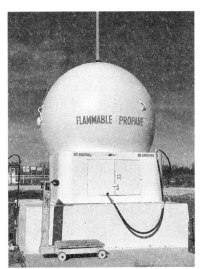

Also, very commonly we see a combination of both.

This tank has a cylindrical body and hemispherical ends.

If we pressurize a tank such as this, the body of the tank will expand due to that pressure.

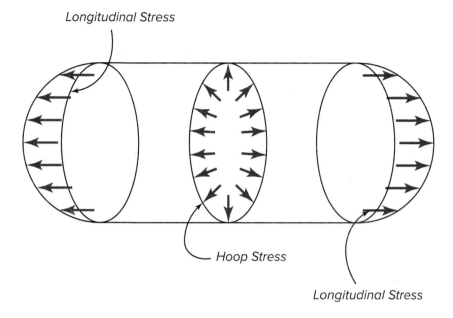

Longitudinal Stress

Hoop Stress

Longitudinal Stress

As that tank is pressurized, it grows in two different directions. One is an expansion of the diameter of the tank, known as hoop stress.

The second is the ends of the tank being being pushed outward, known as the longitudinal stress. We can calculate the stresses in the tank body as follows:

$$\sigma_{hoop} = \frac{Pr}{t}$$

σ_{hoop} = normal stress in vessel walls
P = internal pressure of the tank
r = inside radius of the tank
t = wall thickness of the tank

$$\sigma_{long} = \frac{Pr}{2t}$$

σ_{long} = longitudinal or also called normal stress

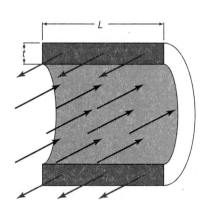

The derivation of these equations helps visualize why the stresses in a spherical pressure vessel (which seems like an infinite number of hoops) is different from a cylindrical vessel.

Equations of equilibrium from the free body diagram:

$$P2rL - \sigma_{hoop}\, 2tL = 0 \quad \rightarrow \quad \sigma_{hoop} = \frac{P2rL}{2tL} = \frac{Pr}{t}$$

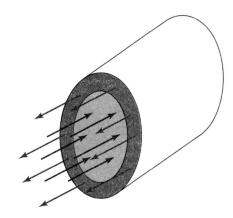

Equation of equilibrium form the free body diagram:

$$P\pi r^2 - \sigma_{long}2\pi rt \;\rightarrow\sigma_{long} = \frac{P\pi r^2}{\pi 2rt} = \frac{Pr}{2t}$$

where: $2\pi r$ is the circumference

Note: The FBD is the same for the longitudinal direction of a pressure vessel as well as a spherical pressure vessel.

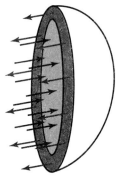

For a spherical pressure vessel, it has longitudinal stress in both of its x and y directions. As a result of the stresses being halved, we typically see spherical tanks housing higher pressured materials than those in cylindrical tanks. This is also why deep sea exploration vehicles are always spherical.

So why isn't there a stress in the z direction (radial stress)? In this direction the stress technically varies from P inside the tank to atmosphere on the outside over a very small distance (the wall thickness). This stress is negligible compared to the axial stress and so the radial stress can be ignored.

So what property makes it a "thin"-walled pressure vessel? Generally, we say that if the ratio of radius to the wall thickness is greater than 10 ($\frac{r}{t}$>10), it is a thin-walled vessel and the equations mentioned above would apply.

Stress Elements

We have discussed stress elements (infinitesimally small material samples that are under load) several times before, but they are going to become even more important in this level.

Even though stress elements are always 3D, we can often simplify them by just looking at one face or a plane and looking at them as 2D elements. In these cases, the stress on the z-face that would be facing you would be zero.

Since an element can have both shear stress and normal stress, they are both included on elements as shown in the example below.

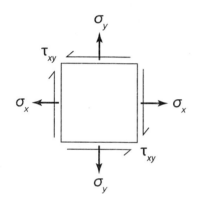

σ_x pointed in the x direction is said to be on the x-face of the stress element.

σ_y pointed in the y direction is said to be on the y-face of the stress element.

τ_{xy} is the shear stress, and its values must be the same on each face always. (Otherwise, the forces would be unbalanced and the stress elements would be spinning!)

In addition, you must understand the orientation and terminology for τ_{xy}. The first letter is the face that the stress is on and the second letter is the direction.

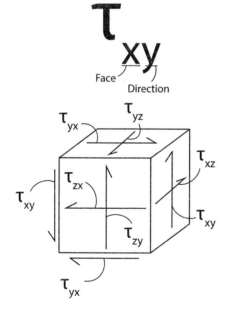

WATCH VIDEO

MECHANICS OF MATERIALS LESSON 42
Thin-Walled Pressure Vessel Example Problem

EXAMPLE: THIN-WALLED PRESSURE VESSELS

Find the hoop stress and longitudinal stress on the cylindrical tank railcar. Its contents are carried at 400 psi. The wall thickness is $\frac{7}{16}$" and the outer diameter is 9.5'.

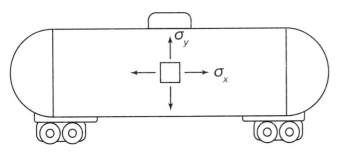

COMBINED LOADING

STEP 1: Calculate hoop stress. Hoop stress is on our stress element.

$$\sigma_{hoop} = \frac{Pr}{t}$$

$$P = 400\frac{\text{lbs}}{\text{in}^2}$$

$$r = (\frac{(9.5 \times 12)}{2}) - \frac{7}{16} = 56.56 \text{ inches}$$

$$t = 0.4375 \text{ inches}$$

$$\sigma_{hoop} = \frac{400(56.56)}{0.4375}$$

$$\sigma_{hoop} = 51,712 \text{ psi or } 51.71 \text{ ksi}$$

STEP 2: Calculate longitudinal stress. Longitudinal stress is σ_x on our stress element.

$$\sigma_{long} = \frac{Pr}{2t}$$

Since longitudinal stress is half as much as hoop stress, we can just divide hoop stress by 2.

$$\sigma_{long} = \frac{51.71}{2} = 25.86 \text{ ksi}$$

COMBINED LOADING

PRO TIP

When solving for the hoop and longitudinal stress in a pressure vessel, the *r* in the equation is the inside radius but honestly it doesn't really matter if you use the inner or outer radius since it is a thin-walled vessel. If you use the outer radius, you are slightly overestimating the stress but overestimating is better than under when it comes to safety!

Press pause on video lesson 42 once you get to the workout problem. Only press play if you get stuck.

Draw a stress element and find all stresses on point *A*. The internal tank pressure is 80 psi. The inner diameter is 22". The wall thickness is 0.25".

TEST YOURSELF 8.1

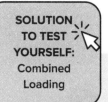

SOLUTION TO TEST YOURSELF: Combined Loading

The cylindrical pressure vessel has an outer diameter of 39 in and a wall thickness of $\frac{7}{32}$ in.

1. For an internal pressure of 250 psi, what is the hoop stress? Is it larger than the longitudinal stress?

2. If the limit stress of the cylinder material is 20 ksi, what is the allowable pressure?

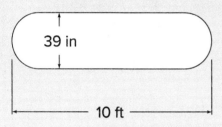

39 in

10 ft

ANSWERS

1. 22,031 psi; yes, that is twice the longitudinal stress

2. $P = 227$ psi

▶ WATCH VIDEO **MECHANICS OF MATERIALS LESSON 43**
Harder Thin-Walled Pressure Vessel Example Problem

▶ Press pause on video lesson 43 once you get to the workout problem. Only press play if you get stuck.

If the 23.99" ring has to fit on the 24" outer diameter barrel and the ambient temp is 70°F, find the temperature to make the ring fit the barrel. Once cooled back to 70°F, find the stress in the ring. Find the pressure in the barrel created by the ring. Assume that the steel barrel has the same pressure throughout.

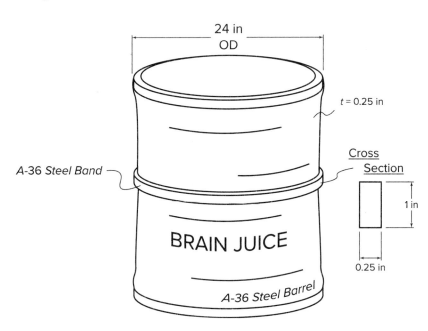

COMBINED LOADING

SOLUTION
TO TEST
YOURSELF:
Combined
Loading

✓ TEST YOURSELF 8.2

A cylindrical pressure vessel of radius *r* and wall thickness *t* is made by wrapping the metal skin at a 45° angle and welding the joint. Express the normal stress perpendicular to the weld in terms of *P*, *r*, and *t*.

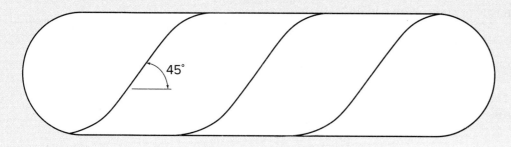

COMBINED LOADING

MECHANICS OF MATERIALS LESSON 44
Combined Loading Introduction Problem

Now that we have *all* of the stress concepts under our belts, we can begin to look at loading scenarios, where all of these stresses are applied to a single body. Recall these stresses:

$$\sigma = \frac{P}{A}$$ (due to tensile or compressive loads)

$$\sigma = \frac{Mc}{I}$$ (stress from beam bending)

$$\sigma = \frac{Pr}{t}, \frac{Pr}{2t}$$ (hoop stress or longitudinal stress from pressure vessels)

$$\tau = \frac{VQ}{It}$$ (transverse shear from bending)

$$\tau = \frac{Tc}{J}$$ (shear stress from torsional loading)

The goal for this level will be to look at a point on a complexly loaded body and determine its state of stress. This is most easily understood after creating a stress element at the point of interest and analyzing all the stress on this point.

COMBINED LOADING

EXAMPLE: COMBINED LOADING INTRO

The "C" clamp has been tightened to apply a force of 500# on a part. Find the stress on parts A and B. Assume that the cross-section is simply rectangular.

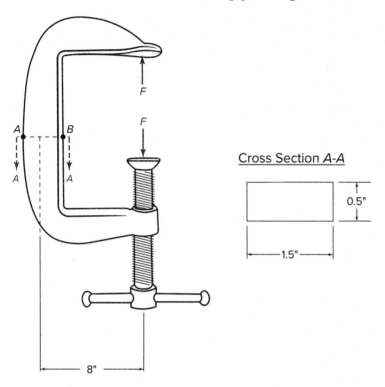

Cross Section *A-A*

0.5"

1.5"

8"

STEP 1: Section the part through the section of interest (in this case, section *A-A*), and draw the FBD of either half of the part.

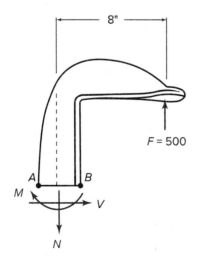

8"

F = 500

A   B

M

V

N

STEP 2: Solve for M, N, and V.

$$\Sigma F_y = 0 = 500 - N$$
$$\therefore N = 500\#$$
$$\Sigma F_x = 0 = V$$
$$\therefore V = 0$$
$$\Sigma M_{cut} = 0 = -M + 500(8)$$
$$M = 4{,}000 \text{ in}\#$$

STEP 3: Calculate I (Moment of Inertia) for the cross section.

Cross Section *A-A*

$$I_{xx} = \frac{1}{12}(0.5)(1.5)^3 = 0.1406 \text{ in}^4$$

STEP 4: Calculate the state of stress for point A.

$$\sigma_A = \frac{P}{A} - \frac{Mc}{I}$$

$\dfrac{P}{A}$ is positive because the cross section is in tension.

$\dfrac{Mc}{I}$ is negative at point A because that point is on the outside of the part and is in compression due to bending.

$$\sigma_A = \frac{500\ \#}{(1.5")(0.5")} - \frac{4{,}000 \text{ in}\#\ (0.75")}{0.1406 \text{ in}^4} = -20{,}670.5 \text{ psi}$$

STEP 5: Calculate the state of stress for point B.

$$\sigma_B = \frac{P}{A} + \frac{Mc}{I}$$

$\dfrac{P}{A}$ is positive because the cross section is in tension.

$\dfrac{Mc}{I}$ is positive because point B is in tension due to bending.

$$\sigma_B = \frac{500\ \#}{(1.5")(0.5")} + \frac{4{,}000 \text{ in}\#\ (0.75")}{0.1406 \text{ in}^4} = 22{,}004 \text{ psi}$$

Press pause on video lesson 44 once you get to the workout problem. Only press play if you get stuck.

If $\sigma_{allowable}$ = 30 ksi, find *Pmax* that can be applied to the bracket. The bracket is made from 1.5" diameter round bar.

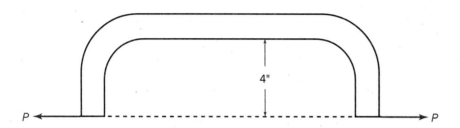

✓ TEST YOURSELF 8.3

SOLUTION TO TEST YOURSELF: Combined Loading

For the bracket shown, what are the maximum tensile and compressive normal stress at the fixed base? The bracket is made from a solid steel 0.25 in diameter rod.

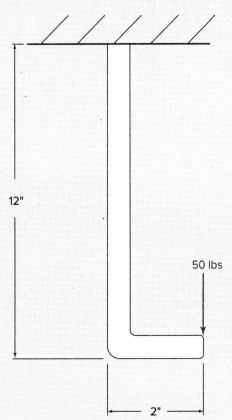

12"

50 lbs

2"

COMBINED LOADING

WATCH VIDEO **MECHANICS OF MATERIALS LESSON 45**
Combined Loading, the Graphical Method

The "Effect Map"

Recall our earlier discussion of the "effect map." This will again be very useful as we analyze loads and their "effects" on the stress elements.

Any given force can create tension or compression at a point ($+F$), bending moment (M), a shearing force (V), or a torsion (T). These "effects" must then be added together to find the total amount of normal stress, and finally, shear stress.

 ## EXAMPLE: COMBINED LOADING

Find the state of stress on point A and draw a volume element cube.

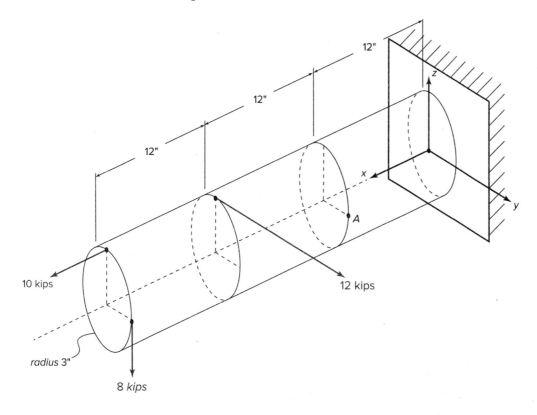

COMBINED LOADING

STEP 1: Draw the cross section containing point A.

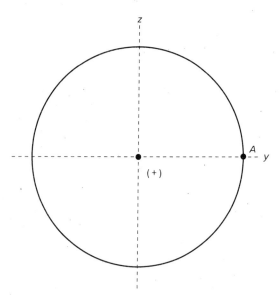

STEP 2: Draw the "effects" from each force on the cross section from above. Let's start with the 10 kip load.

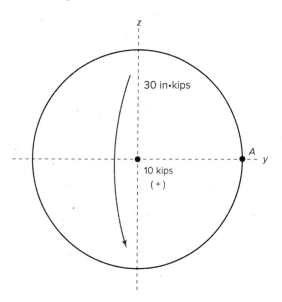

1. The 10 kip load is causing a tensile ($\frac{P}{A}$) stress. (Shown as a dot with a positive sign [tension] right in the center.)
2. Since the 10 kip load is not on the neutral axis of the beam, it is causing a small bending moment. $M = 10$ kip $\times 3" = 30$ in·kips
3. This load is not causing any shear (V) or torsion (T).

STEP 3: Let's add the effects from the 8 kip load. Add these effects to the effect map from step 2.

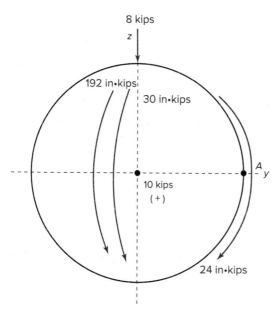

1. The 8 kip load causes no tension or compression on the cross section containing point A.

2. The 8 kip load is causing a bending moment at cross section containing point A.
   $M = 8\,\text{kip} \times 24" = 192\,\text{in·kips}$

3. The 8 kip load is causing a shear force (V) across the section containing point A. (Shown as an arrow in the direction of the shear.)

4. Finally, the 8 kip load is causing torsion. $T = 3" \times 8\,\text{kips} = 24\,\text{in·kips}$ (Show as torque arrow.)

STEP 4: Finally, let's add in the effect from the 12 kip load and add this to our map.

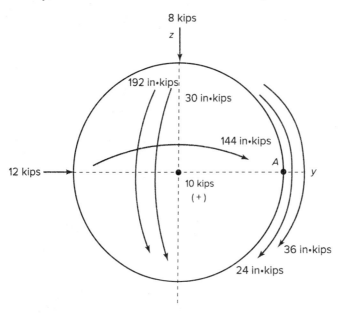

1. The 12 kip load does not add any tension or compression stress to the section containing point A.
2. The 12 kip load does make a bending moment.
   $A = 12 \text{ kips} \times 12" = 144 \text{ in·kips}$.
3. The 12 kip load also causes a shear force (V) on the plane. (Shown as a shear force arrow.)
4. Finally, the 12 kip force causes a torsion to the shaft.
   $T = 12 \text{ kips} \times 3" = 36 \text{ in·kips}$.

STEP 5: Simplify our "effect map" by combining like forces, moments, and torques.

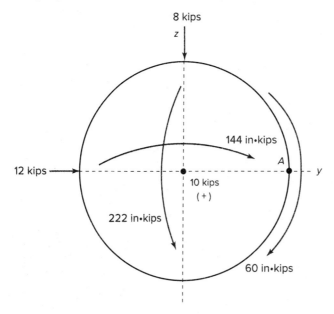

STEP 6: Let's calculate the normal stress (σ) on point A.

When bending about the y-axis, point A would be on the neutral axis, so the 222 in·kips produces no stress at point A.

The 144 in·kip moment bending about the z-axis would cause bending stress and cause point A to be in compression ($-\sigma$).

Finally, the tensile force in the center of the beam would cause positive normal stress ($\frac{P}{A}$).

$$\sigma_A = \frac{P}{A} - \frac{Mc}{I} = \frac{10\ \text{kips}}{\pi(3)^2\ \text{in}^2} - \frac{(144\ \text{in kips})(3")}{\frac{\pi}{4}(3)^4\ \text{in}^4}$$

$$\sigma_A = 0.354\ \text{ksi} - 6.79\ \text{ksi} = -6.43\ \text{ksi}$$

STEP 7: Calculate shear stress (τ) on the section.

Recall transverse shear is max on the neutral axis and zero on the outside of the part.

1. The 12 kip shear force will produce zero transverse shear (τ) since point A would be on the bottom of the part. (Rotate your perspective 90°.)
2. The 8 kip force would produce transverse shear as point A would be on the neutral axis.

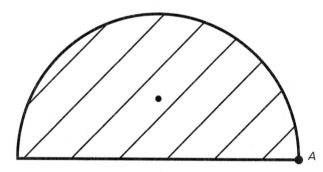

$$Q = y_A = \left(\frac{4(3)}{3\pi}\right)\left(\frac{\pi(3)^2}{2}\right) = 18\ \text{in}^3$$

Let's look only at those things causing shear stress.

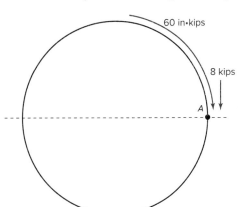

If I draw these effects on point *A* as shown, it is easy to see that they both act on point *A* in the same direction. Therefore, they will add together!

$$\tau_A = \frac{VQ}{It} + \frac{Tc}{J}$$

$$\tau_A = \frac{8 \text{ kip}(18\text{in}^3)}{\frac{\pi}{4}(3)^4 \text{ in}^4(6")} + \frac{(60 \text{ in kips})(3")}{\frac{\pi}{2}(3)^4 \text{ in}^4} = 0.377 \text{ ksi} + 1.414 \text{ ksi}$$

$$\tau_A = 1.79 \text{ ksi}$$

STEP 8: Finally, let's construct our volume element. Here is an easy way to get your shear stress directions correct.

Think of point *A* as a piece of material that would lie between two halves of the beam. Since the shear forces from the left portion of the beam (where all the loads were) are downward, simply draw them, and the rest has to alternate up and down!

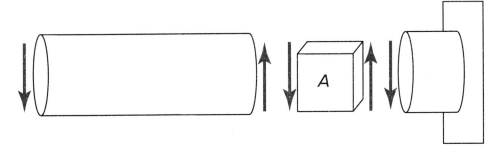

★ **Key Point:** The signs on the shear stress can be tricky. Follow this method and you can't go wrong!

COMBINED LOADING

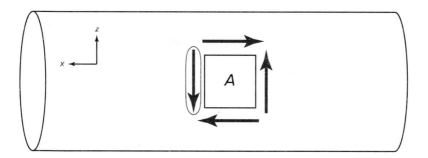

Now we will have the correct directions on our final model.

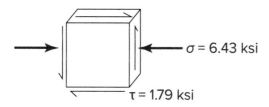

$\sigma = 6.43$ ksi

$\tau = 1.79$ ksi

▶ Press pause on video lesson 45 once you get to the workout problem. Only press play if you get stuck.

Determine the state of stress at points *A* nd *B* that has a 50 mm diameter. Then draw the resulting stress element.

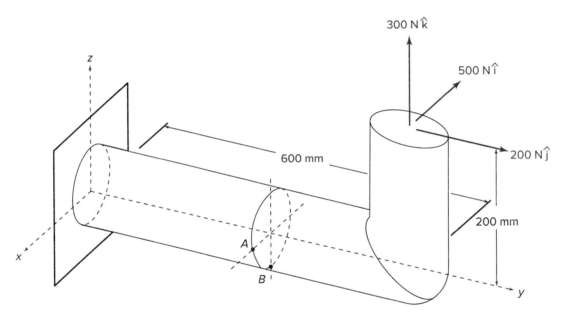

✓ TEST YOURSELF 8.4

SOLUTION
TO TEST
YOURSELF:
Combined
Loading

For the circular cross-section L-bracket shown, compute the normal and shear stress at points 1 and 2 at a distance 8 ft from the end (as shown). The rod has a diameter of 2 in.

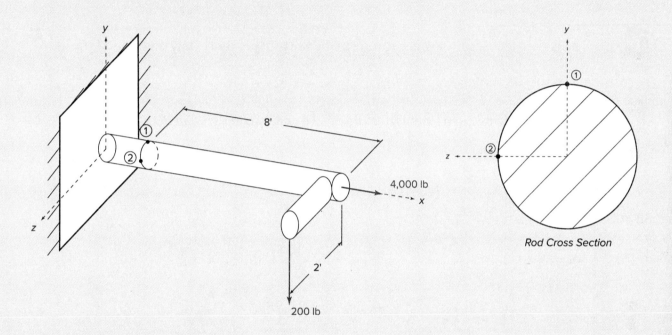

Rod Cross Section

COMBINED LOADING

There is nothing really new regarding the content for this problem. It simply has more "stuff" to keep track of (i.e., a combination of everything you have learned to this point).

▶ Press pause on video lesson 46 once you get to the workout problem. Only press play if you get stuck.

For the bevel gear with the given loading, find the total σ and τ acting at point *B*. Let's do point *A* also.

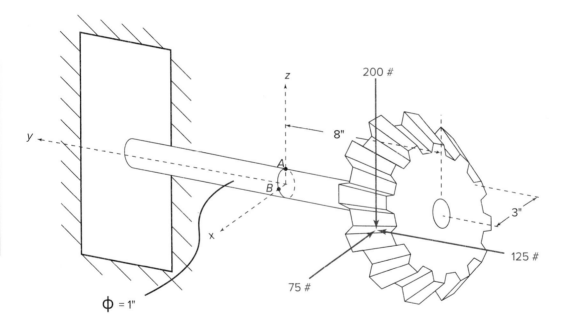

COMBINED LOADING

MECHANICS OF MATERIALS LESSON 47
Combined Loading Continued

EXAMPLE: COMBINED LOAD ON A BEAM

Find the state of stress on points A and B on the beam below and draw stress elements for each. The cross section of the beam is also given below.

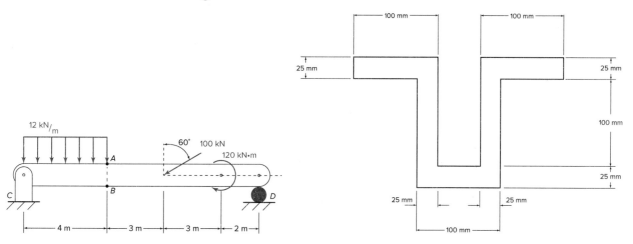

STEP 1: Find global equilibrium of the beam.

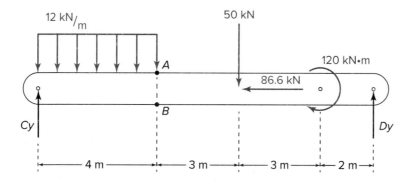

$$\Sigma M_c = 0 = -48(2) - 50(7) - 120 + Dy(12)$$
$$Dy = 47.17 \text{ kN}$$

$$\Sigma Fy = 0 = Cy + 47.17 - 48 - 50$$
$$Cy = 50.83 \text{ kN}$$

COMBINED LOADING

STEP 2: Section the beam through the plane containing points A and B, draw that FBD, and solve for M, N, and V.

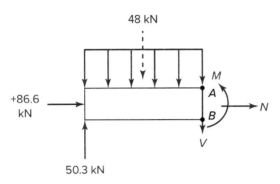

$$\Sigma F_x = 0 = 86.6 + N$$
$$\therefore N = -86.6 \text{ kN}$$

$$\Sigma F_y = 0 = -V + 50.83 - 48$$
$$\therefore V = 2.83 \text{ kN}$$

$$\Sigma M_{cut} = 0 = M + 48(2) - 50.83(4)$$
$$M = 107.32 \text{ kN}$$

STEP 3: Calculate $\bar{y}$ for the cross section in order to find the neutral axis of the beam.

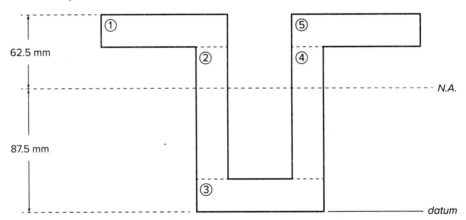

Area	y_i	A_i	y_iA_i
1	137.5 mm	2,500 mm²	343,750 mm³
2	75 mm	2,500 mm²	187,500 mm³
3	12.5 mm	2,500 mm²	31,250 mm³
4	75 mm	2,500 mm²	187,500 mm³
5	137.5 mm	2,500 mm²	343,750 mm³
		$\Sigma A = 12,500$ mm²	$\Sigma y_iA_i = 1,093,750$ mm³

$$\bar{y} = \frac{1,093,750}{12,500} = 87.5 \text{ mm}$$

STEP 4: Calculate I, the second area moment of inertia.

$$I_{xx} = 2[\frac{1}{12}(100)(25)^3 + (2{,}500)(50)^2] + 2[\frac{1}{12}(25)(100)^3 + (2{,}500)(12.5)^2] + \frac{1}{12}(100)(25)^3$$
$$+ (2{,}500)(75)^2$$
$$I_{xx} = 12{,}760{,}416.67 + 4{,}947{,}916.67 + 14{,}192{,}708.33$$
$$= 31{,}901{,}041.67 \text{ mm}^4$$

STEP 5: Since there is no torsion on the beam, and points A and B are on the top and bottom fibers of the beam respectively, transverse shear stress will be zero since Q will be zero at both of these points.

$\therefore$ there is no shear stress at points A and B!

STEP 6: Calculate normal stress at points A and B. Due to the moment on the beam, point A (top) would be in compression, and point B (bottom) would be in tension. Also, both A and B would be in compression due to the 86.6 kN normal force.

$$\sigma_A = -\frac{P}{A} - \frac{Mc}{I} = \frac{-86{,}600 \text{ N}}{12{,}500 \text{ mm}^2} - \frac{(107{,}320{,}000 \text{ N•mm})(62.5 \text{ mm})}{31{,}901{,}041.67 \text{ mm}^4}$$

$$\sigma_A = -6.928 \text{ MPa} - 210.26 \text{ MPa} = -217.19 \text{ MPa}$$

$$\sigma_B = -\frac{P}{A} + \frac{Mc}{I} = \frac{-86{,}600 \text{ N}}{12{,}500 \text{ mm}^2} + \frac{(107{,}320{,}000 \text{ N•mm})(87.5 \text{ mm})}{31{,}901{,}041.67 \text{ mm}^4}$$

$$\sigma_B = -6.928 + 294.36 \text{ MPa} = 287.44 \text{ MPa}$$

STEP 7: Finally, construct stress elements for each.

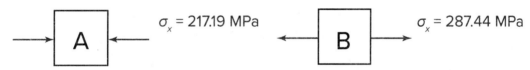

$\sigma_x = 217.19$ MPa Compression
$\sigma_x = 287.44$ MPa Tension

Press pause on video lesson 47 once you get to the workout problem. Only press play if you get stuck.

Determine the state of stress at points *A* and *B*. Then draw the resulting stress element.

Note: The pins at *C* and *D* are on the neutral axis.

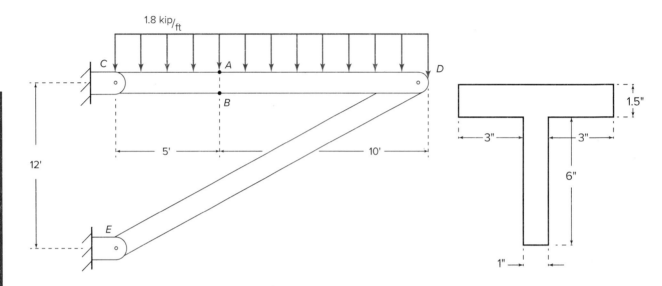

TEST YOURSELF 8.5

SOLUTION TO TEST YOURSELF: Combined Loading

A cantilevered beam is loaded by a cable attached to a 1,000 lb block via a frictionless pulley as shown. The cross section of the beam is also shown below. It has a thickness of $\frac{1}{8}"$ everywhere.

1. Compute the largest tensile and compressive normal stress in the beam.
2. What is the shear stress at the neutral axis?

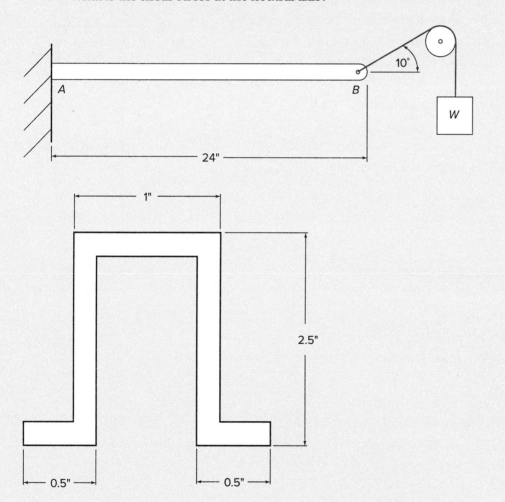

ANSWERS

$\sigma = 10{,}045$ psi (Tension)

$\sigma = 7{,}621$ (Compression)

$\tau = 361$ psi

Thin-Walled Pressure Vessels

- If we pressurize a tank such as this, the body of the tank will expand due to that pressure.
- For a cylindrical vessel it has both hoop stress and longitudinal stress.
- For a spherical vessel it has longitudinal stress in both directions (x and y).

$$\sigma_{hoop} = \frac{Pr}{t}$$

σ_{hoop} = normal stress in vessel walls
P = internal pressure of the tank
r = inside radius of the tank
t = wall thickness of the tank

$$\sigma_{long} = \frac{Pr}{2t}$$

σ_{long} = longitudinal or also called normal stress

- For a vessel to be considered thin-walled and use the equation above, ratio of radius to the wall thickness should be greater than 10, i.e., $\frac{r}{t} > 10$.

Combined Loading

- Any force can produce several "effects" such as:
  - Tension/compression, $\sigma = \dfrac{P}{A}$
  - Transverse shear, $\tau = \dfrac{VQ}{It}$
  - When it is at a distance, a bending moment $\sigma = \dfrac{Mc}{I}$ or a torsion, $\tau = \dfrac{Tc}{J}$

The "Effect Map"

Any given force can create tension or compression at a point ($\pm F$), bending moment (M), a shearing force (V), or a torsion (T). These "effects" must then be added together to find the total amount of normal stress and, finally, shear stress.

Equations Learned in This Level

Hoop Stress Formula

$$\sigma_{hoop} = \frac{Pr}{t}$$

Longitudinal Stress Formula

$$\sigma_{long} = \frac{Pr}{2t}$$

PRO TIP

Pressure Vessel
- When solving for the hoop and longitudinal stress in a pressure vessel, the *r* in the equation is the inside radius but honestly it doesn't really matter if you use the inner or outer radius since it is a thin-walled vessel. If you use the outer radius, you are slightly overestimating the stress but overestimating is better than under when it comes to safety!

Level 9
Stress Transformation

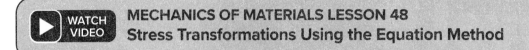

Stress Transformations

Stress is directional. If you only look at stress on a body at an *x, y* orientation, you may not be looking at the maximum state of stress. In this case, we must rotate the stress element and find the maximum normal and shear stresses. This would become important if we wanted to find the stress along a weld seam, for instance, as in the example below.

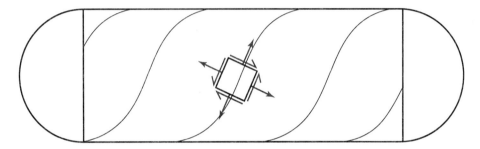

We could find this stress element on the tank above by finding an *x, y* stress element which would be derived by finding hoop stress and longitudinal stress, resulting in an element like this:

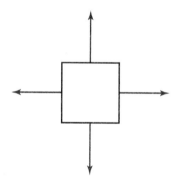

Note, in this case, τ_{xy} (shear stress) is zero.

STRESS TRANSFORMATION

Now we can rotate this element until we arrive at the element desired.

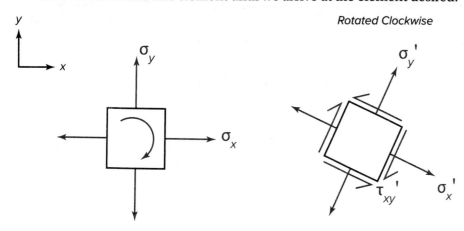

Rotated Clockwise

Notice now we have shear stress. Where did that come from? Think of the original σ_x. As we rotate σ_x, it can be broken into components. One component becomes σ_x', and the other component becomes τ_{xy}'.

For stress transformations, we have several equations to find special stress cases.

Stress Transformation Equations

$$\sigma_x' = \frac{\sigma_x + \sigma_y}{2} + \frac{\sigma_x - \sigma_y}{2}\cos(2\theta) + \tau_{xy}\sin(2\theta)$$

$$\sigma_y' = \frac{\sigma_x + \sigma_y}{2} - \frac{\sigma_x - \sigma_y}{2}\cos(2\theta) - \tau_{xy}\sin(2\theta)$$

$$\tau_{xy}' = -\left(\frac{\sigma_x - \sigma_y}{2}\right)\sin(2\theta) + \tau_{xy}\cos(2\theta)$$

In these equations, θ is the angle between the x and x' axes.

Note: In some books authors like to use σ_n (normal) instead σ_x' and σ_t (tangential) instead of σ_y'.

To use these equations, you will need a stress on the original element that you will then transform to an angle. In addition to the element, you will need the transformed angle desired as well as the direction of rotation (clockwise or counterclockwise).

PRO TIP

Make sure to pay attention to the sign on the angle when using the stress transformation equations. If the angle is clockwise as measured from the x to x' axis, then the angle is negative. If the angle is counterclockwise as measured from the x to x' axis, then the angle is positive.

STRESS TRANSFORMATION

EXAMPLE: STRESS TRANSFORMATION EQS

Rotate the given element 27° counterclockwise and find the transformed stress. Draw the new stress element.

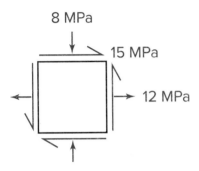

8 MPa

15 MPa

12 MPa

STEP 1: Identify the following items from the given stress element.

$\sigma_x = 12$ MPa (Tension is positive.)

$\sigma_y = -8$ MPa (Compression is negative.)

$\tau_{xy} = 15$ MPa

(The easy way to determine the sign on the shear is to simply look at the shear stress on the x-face and determine if that stress causes the element to rotate clockwise or counterclockwise. Counterclockwise is positive; clockwise is negative.)

$\theta = 27°$

STEP 2: Calculate σ_x'.

$$\sigma_x' = \frac{\sigma_x + \sigma_y}{2} + \frac{\sigma_x - \sigma_y}{2} \cos(2\theta) + \tau_{xy} \sin(2\theta)$$

$$\sigma_x' = \frac{12 + (-8)}{2} + \frac{12 - (-8)}{2} \cos((2)(27°)) + 15 \sin((2)(27°))$$

$$\sigma_x' = 2 + 5.88 + 12.13 = 20.01 \text{ MPa}$$

STEP 3: Calculate σ_y'.

$$\sigma_y' = \frac{\sigma_x + \sigma_y}{2} - \frac{\sigma_x - \sigma_y}{2} \cos(2\theta) - \tau_{xy} \sin(2\theta)$$

$$\sigma_y' = \frac{12 + (-8)}{2} - \frac{12 - (-8)}{2} \cos((2)(27°)) - 15 \sin((2)(27°))$$

$$\sigma_y' = 2 - 5.88 - 12.14 = -16.02 \text{ MPa}$$

STRESS TRANSFORMATION

STEP 4: Calculate τ_{xy}'.

$$\tau_{xy}' = -(\frac{\sigma_x - \sigma_y}{2})sin(2\theta) + \tau_{xy}cos(2\theta)$$

$$\tau_{xy}' = -(\frac{12 - (-8)}{2})sin((2)(27°)) + 15\,cos((2)(27°))$$

$$\tau_{xy}' = -8.09 + 8.82 = 0\,.727\ \text{MPa}$$

STEP 5: Draw the new stress element at 27° counterclockwise.

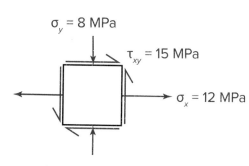

Original Stress Element

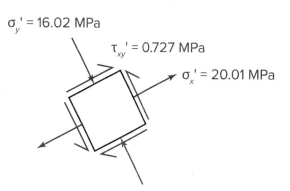

Rotated Element 27° Counterclockwise

PRO TIP

When using the equations to calculate stress transformations, be super careful with your signs as this will change your answer. Calculating using Mohr's Circle (which will be covered later) will give you a better visualization of your answers and reduce mistakes.

STRESS TRANSFORMATION

▶ Press pause on video lesson 48 once you get to the workout problem. Only press play if you get stuck.

Find σ_x', σ_y', and τ_{xy}' for the stress element after it is rotated 60° clockwise.

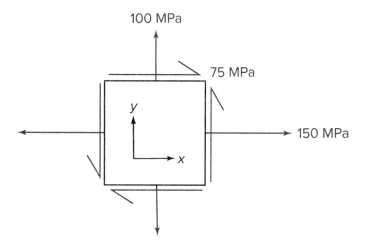

TEST YOURSELF 9.1

SOLUTION TO TEST YOURSELF: Stress Transformation

With $\sigma_x = 4{,}000$ psi, $\sigma_y = 0$, and $\tau_{xy} = -500$ psi, what are the stresses if the x-axis is rotated 30° downward (clockwise)?

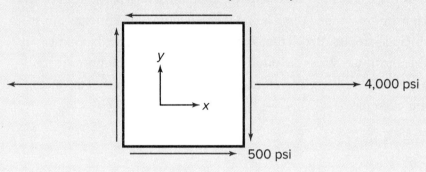

4,000 psi

500 psi

ANSWERS

$\sigma_x{}' = 3{,}433$ psi
$\sigma_y{}' = 567$ psi
$\tau_{xy}{}' = 1{,}482$ psi

MECHANICS OF MATERIALS LESSON 49
Max Shear and Principal Stress with Equation Method

Max Shear and Principal Stresses

If stress is directional and can vary as you rotate the stress element, there must be a particular position where the normal stress becomes maximum. Indeed there is and it's called the Max Principal Stress. There are actually two in-plane principal stresses. Recall that stress elements have an x-face and a y-face, so when the one face is at the maximum stress the other face will be at the minimum stress. Both of these values are known as Principal Stresses. Therefore, σ_1 will be acting on one of the faces, and σ_2 will be acting on the perpendicular face.

$$\sigma_1, \sigma_2 = \frac{\sigma_x + \sigma_y}{2} \pm \sqrt{\left(\frac{\sigma_x - \sigma_y}{2}\right)^2 + \tau_{xy}^2}$$

Conversely, there's a position when the stress element is rotated where shear stress becomes maximum, and that's called the Max Shear.

$$\tau_{xy\,max} = \pm \sqrt{\left(\frac{\sigma_x - \sigma_y}{2}\right)^2 + \tau_{xy}^2}$$

FUN FACTS

When the element is rotated so that the faces are aligned with the principal stresses, shear stress will *always* be zero. In addition, when the element is rotated to the maximum shear stress state, the normal stress will *always* equal the average stress of the element (the stress on the x-face will be exactly the same as the y-face).

STRESS TRANSFORMATION

Principal Stress Angles

The principal stress angles are the angles between the given state of stress and how much that stress element needs to rotate to reach the principal stress. Since the element could be rotated clockwise or counterclockwise to reach principal stresses, there will be two principal stress angles, given as θ_{p_1} and θ_{p_2}.

θ_{p_1} will always be the smallest amount of rotation to reach a principal stress, and θ_{p_2} will be the largest. The signs for these angles depend upon whether you have to rotate clockwise or counterclockwise to reach the principal stresses. (Clockwise is negative, and counterclockwise is positive.)

$$tan(2\theta_p) = \frac{\tau_{xy}}{(\frac{\sigma_x - \sigma_y}{2})}$$

Maximum Shear Stress Angles

Just like the principal stress angles, there is an angle to which you can rotate your stress element to reach the maximum shear stress value. This angle is called the maximum shear stress angle (θ_s).

Since the element could be rotated clockwise or counterclockwise to reach maximum shear stresses, there will be two maximum shear stress angles, given as θ_{s_1} and θ_{s_2}.

θ_{s_1} will always be the smallest amount of rotation to reach a maximum shear stress, and θ_{s_2} will be the largest. The signs for these angles depend upon whether you have to rotate clockwise or counterclockwise to reach the maximum shear stresses. (Clockwise is negative, and counterclockwise is positive.)

$$tan(2\theta_s) = -\left[\frac{(\frac{\sigma_x - \sigma_y}{2})}{\tau_{xy}}\right]$$

PRO TIP

The magnitude of the principal stress angles (θ_{p_1} and θ_{p_2}) will always add to 90°.

Likewise, the magnitude of the maximum shear stress angles (θ_{s_1} and θ_{s_2}) will also always add to 90°.

STRESS TRANSFORMATION

▶ Press pause on video lesson 49 once you get to the workout problem. Only press play if you get stuck.

F or the following stress element:

1. Find the principal stress.
2. Find the max in-plane shear stress.
3. Specify the orientation for parts 1 and 2.

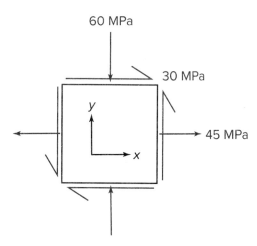

✓ TEST YOURSELF 9.2

SOLUTION TO TEST YOURSELF: Stress Transformation

T ake $\sigma_x = 45$ MPa, $\sigma_y = -18$ MPa, and $\tau_{xy} = 9$ MPa. What are:

1. The principal stresses and the rotation angle θ_p
2. The maximum in-plane shear stress and the rotation angle θ_s

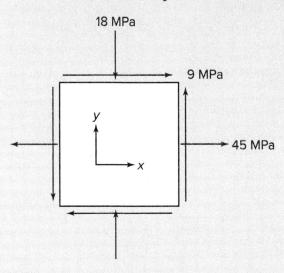

18 MPa

9 MPa

y

45 MPa

x

ANSWERS

$\sigma_1 = 46.26$ MPa

$\sigma_2 = -19.26$ MPa

$\theta_p = 7.97°$ (or 97.97°)

$\tau_{xy\,max} = 32.76$ MPa

$\theta_s = -37.03°$ (or 52.97°)

STRESS TRANSFORMATION

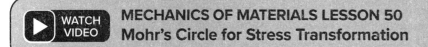

MECHANICS OF MATERIALS LESSON 50
Mohr's Circle for Stress Transformation

Mohr's Circle

Mohr's Circle is a 2D geometric representation of stress transformation in a stress element. The circle is plotted using normal and shear stresses as coordinate points.

Once you have learned this technique, hopefully, you will never go back to using the equation method to calculate stress transformations.

The first thing you have to understand for Mohr's Circle is the coordinate system.

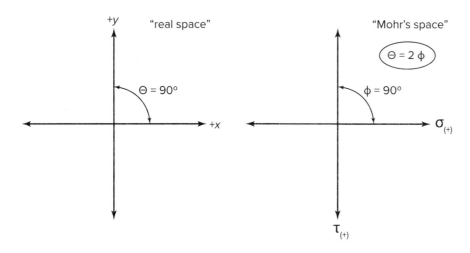

- **Point 1.** Notice that $\tau_{(+)}$ is downward. This is counter to what we generally think of as the positive direction.
- **Point 2.** The angles are different in "Mohr's space" than in "real space." To take an angle from the "real world" (usually the angle given in the problem statement to be transformed) into "Mohr's Circle world," you have to multiply the angle by 2.

Likewise, to go the opposite direction, "Mohr's Circle world" to "real world," you would divide the angle by 2.

PITFALL

Your final transformation angle should *always* be reported in "real world" degrees, not in "Mohr's Circle world" degrees.

STRESS TRANSFORMATION

How to Generate Mohr's Circle Coordinates

The first step in constructing Mohr's Circle is always generating a set of coordinates to plot our circle. The values for this come from the stress element.

Typically, you will be given a stress element such as the one seen next:

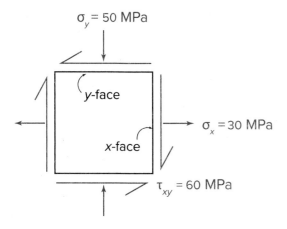

For this stress element, the coordinate points will be easy to produce:
$\tau_{xy} = -60$, $\sigma_y = -50$, and $\sigma_x = +30$

Plugging in values from our stress element:

- **Point 1.** x-face $(\sigma_x, \tau_{xy}) = (30, -60)$
- **Point 2.** y-face $(\sigma_y, -\tau_{xy}) = (-50, -(-60)) = (-50, 60)$

Note: The x-face was in tension $(+)$ and the y-face was in compression $(-)$. Also pay close attention to the arrow direction for the shear stress on the x-face. Shear stress was downward (which would rotate the element clockwise), which is the negative direction. On the y-face, the shear stress rotates the element counterclockwise, which is positive.

PRO TIP

To create Mohr's circle, the key is simply to flip the sign on shear for one of the points that you plot (it doesn't matter which one). This allows you to create a circle that is centered on the x-axis.

Then, make sure you pay attention to what shear you paired with your σ_x and σ_y coordinate axes. Different instructors use different signs conventions, so be consistent (especially pay attention to when you are rotating your stress element).

EXAMPLE: MOHR'S CIRCLE STRESS TRANSFORMATION

Let's use the stress element in the previous problem and find the following things:

1. Maximum in-plane shear stress (τ_{max})
2. Average normal stress
3. The principal stresses (σ_1, σ_2)
4. The principal stress angles (θ_{p_1}, θ_{p_2})
5. Max shear stress angles (θ_{s_1}, θ_{s_2})
6. State of stress if transformed 27° counterclockwise

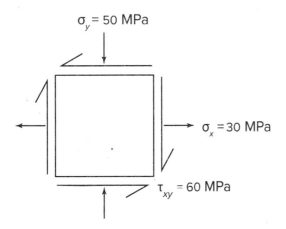

STEP 1:   Generate coordinates from the stress element.

x-face (σ_x, τ_{xy}) = x-face (30, −60)

y-face (σ_y, $-\tau_{xy}$) = y-face (−50, 60)

Note: The magnitude of τ_{xy} for both points is always the same but with opposite signs. *Always!*

STEP 2: Plot the values in "Mohr's space." Be sure to remember that positive is in the downward direction (at least that is how we are going to teach it in this book)!

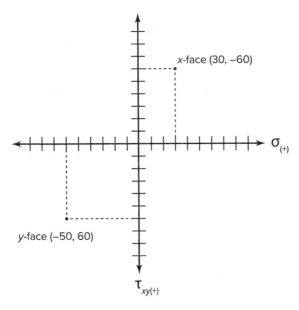

STEP 3: Connect the dots with a straightedge, and then locate the σ value of the center of the line.

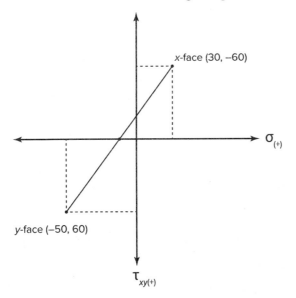

$30 + 50 = 80$	Find the horizontal distance between points.
$\dfrac{80}{2} = 40$	Divide by 2 to find the center.
$-50 + 40 = -10 \ or \ 30 - 40 = -10$	Add to either point to find the center.

(***Note:*** This is the first term in our stress transformation equations.)

A faster way is $(\sigma_x + \sigma_y)/2 = (30 - 50)/2 = -10$

(Note how you keep the sign of each of the stresses and then divide by 2.)

STEP 4: Guess what? We just found the average normal stress!

Average (in the middle) normal stress is the center of the circle.

$\sigma_{\text{AVG}} = -10 \text{ MPa}$

STEP 5: Let's find the principal stresses. The principal stresses are the max and min circle points on the σ axis (horizontal axis). σ_1 is always the rightmost point, and σ_2 is always the leftmost point after constructing our circle through our two coordinate points.

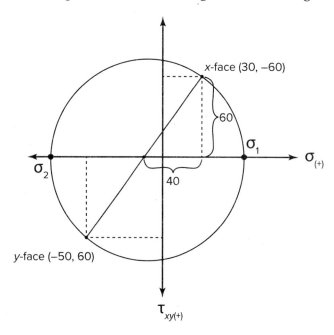

Notice σ_1 is the center of the circle **plus** a radius, while σ_2 is the center of the circle **minus** the radius.

$\therefore$ We need the radius.

$r = \sqrt{(40^2 + 60^2)} = 72.11$ (This is the exact same thing underneath the square root symbol in our stress transformation equations—cool beans!)

The principal stresses are $-10 + 72.11$, so $\sigma_1 = 62.11 \text{ MPa}$ and $\sigma_2 = -82.11 \text{ MPa}$.

Note: A common error is to incorrectly find the radius by taking the horizontal distance to the origin instead of the center of the circle (in the above case accidentally using 30 instead of 40).

PRO TIP

A straightedge and a circle drawing tool such as a compass are a must for generating good drawings and enhancing your understanding.

STRESS TRANSFORMATION

STEP 6: Once again, we get an answer for free!

    If we want to find max in-plane shear stress $(\tau_{xy\,max})$, we simply need to find the value at the topmost or bottommost point of a circle. Since the center of the circle *always* lies on the σ axis, $\tau_{xy\,max}$ is simply one radius away!

$$\tau_{xy\,max} = -72.11 \text{ MPa}$$

STEP 7: Next, let's find the principal stress angles $(\theta_{p_1}, \theta_{p_2})$. This is the angle that we must rotate the current element to reach the principal stresses.

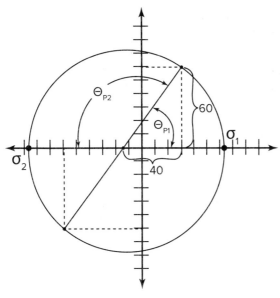

$$tan(\theta_{p_1}) = \frac{60}{40}$$

$\theta_{p_1} = 56.3°$ (***Note:*** This is the angle in "Mohr's Circle World.")

Actual $\theta_{p_1} = \dfrac{56.3°}{2°} = -28.15°$

(Recall the negative is present as we had to rotate clockwise to reach principal stress.)

$\theta_{p_2} = 90° - 28.15° = 61.85°$ (counterclockwise)

STEP 8: Let's now find the maximum shear stress angles $(\theta_{s_1}, \theta_{s_2})$. This stress is extremely similar to the previous step, the only difference being we are now rotating to a vertical axis.

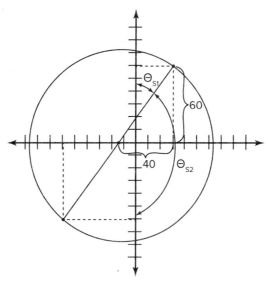

$$tan(\theta_{s_1}) = \frac{40}{60}$$

$\theta_{s_1} = 33.69°$ ("Mohr's Circle World")

$\theta_{s_1} = \dfrac{33.69°}{2} = 16.84°$ (counterclockwse)

$\theta_{s_2} = -90° - 16.84° = -73.16°$ (clockwise)

STEP 9: Finally, let's complete the last step and rotate our stress element 27° counterclockwise (ccw). **Note:** We are rotating 27° in the "real world" so we need to multiply by 2 and rotate 54° in "Mohr's Circle World."

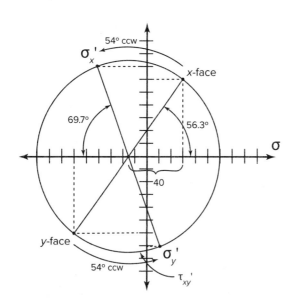

STEP 10: Lastly, calculate the new values for σ_x', σ_y', τ_{xy}'.

Note: $180° - 56.3° - 54° = 69.7°$

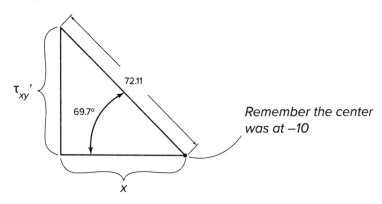

$sin(69.7°) = \dfrac{\tau_{xy}'}{72.11}$, $\tau_{xy}' = 67.63$ MPa

(*Note:* The x-face is in the negative part of the graph and the y-face is in the positive portion.)

$cos(69.7°) = \dfrac{x}{72.11}$, $x = 25.01$ MPa

(*Note:* We need to add this value to the center point for σ_x' and σ_y'.)

$\sigma_x' = -10 - 25.01 = -35.01$ MPa (compression)
$\sigma_y' = -10 + 25.01 = 15.01$ MPa (tension)

Finally, construct the new stress element.

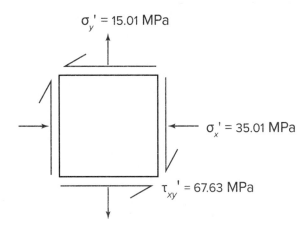

▶ Press pause on video lesson 50 once you get to the workout problem. Only press play if you get stuck.

F or the following stress element, find (1) the average normal stress, (2) the principal stresses (σ_{p_1} and σ_{p_2}), (3) in-plane max shear stress ($\tau_{xy\,max}$), (4) angle to reach the principal stress, (5) σ_x', σ_y', and τ_{xy}' for the stress element after it is rotated 23° clockwise.

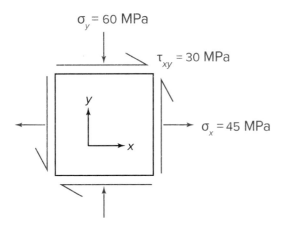

σ_y = 60 MPa

τ_xy = 30 MPa

y

σ_x = 45 MPa

x

✓ TEST YOURSELF 9.3

SOLUTION TO TEST YOURSELF: Stress Transformation

For the stress element shown, use Mohr's Circle to find (1) the average normal stress, (2) the principal stresses (σ_1 and σ_2), (3) in-plane max shear stress ($\tau_{xy\,max}$), (4) angle to reach the principal stress, (5) σ_x', σ_y', and τ_{xy}' for the stress element after it is rotated 8° clockwise. For fun, try solving this problem using a purely graphical approach by accurately drawing Mohr's Circle (yes, use a compass, ruler, and protractor) as well as by the technique shown in video lesson 50.

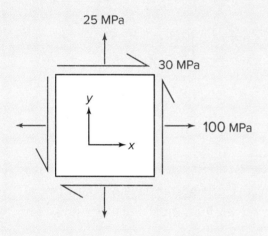

ANSWERS

1. $\sigma_{avg} = 63$ MPa
2. $\sigma_1 = 110$ MPa, $\sigma_2 = 14.5$ MPa
3. $\tau_{xy\,max} = 48$ MPa
4. $\theta_p = 19.35°$ clockwise
5. $\sigma_x' = 90$ MPa, $\sigma_y' = 35$ MPa, $\tau_{xy}' = -40$ MPa

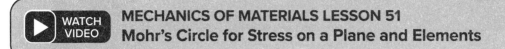

WATCH VIDEO **MECHANICS OF MATERIALS LESSON 51**
Mohr's Circle for Stress on a Plane and Elements

Mohr's Circle Stress on a Plane

Stress transformation is a little bit different than the typical "stress element" transformations that we have been working with if you are asked to find the stress and the shear stress acting on an inclined plane, as in the example next.

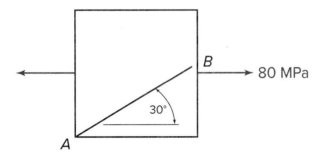

You will only need to solve for normal stress and shear stress on that single plane. These problems can be solved using the stress transformation equations or Mohr's Circle by transforming the x-face of the stress element to the given inclined plane and then simply solving for normal and shear stresses on that x-face (so σ_x' and τ_{xy}').

EXAMPLE: MOHR'S CIRCLE STRESS ON A PLANE

Calculate stresses on plane *AB* on the given stress element below.

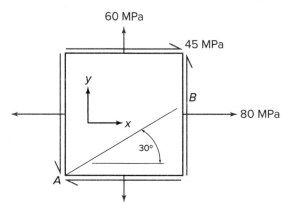

STEP 1: Obtain Mohr's Circle coordinates from the stress element.

x-face $(\sigma_x, \tau_{xy}) = x$-face $(80, 45)$

y-face $(\sigma_y, -\tau_{xy}) = y$-face $(60, -45)$

STEP 2: Plot coordinates from Step 1 on Mohr's Circle.

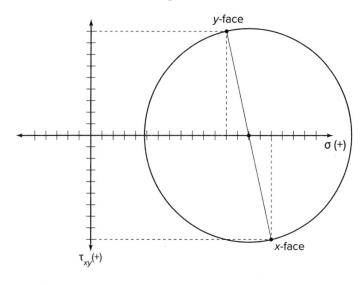

STEP 3: Find the center of the circle (average normal stress) as well as the radius of the circle $(\tau_{xy\,max})$.

$$\text{Center of circle} = 60 + (\frac{80 - 60}{2}) \text{ or } \frac{80 + 60}{2} = 70 \text{ MPa} \qquad \text{(average stress)}$$

$$\text{Radius of circle} = \sqrt{10^2 + 45^2} = 46.1 \text{ MPa} \qquad (\tau_{xy\,max})$$

STEP 4: Next, we need to transform through some angle that will take the *x*-face to the desired plane angle.

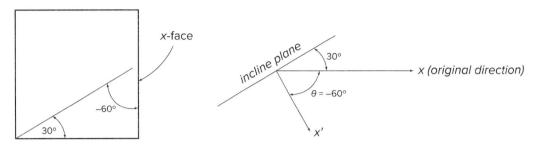

You can either think of the angle as the value needed to rotate the *x*-face to the new *x'*-face (image 1), or you can think of it as rotating the *x*-axis to the *x'*-axis (image 2). In both cases, you will get −60°.

Why negative? Because we have to rotate clockwise to align the original stress element with the desired new plane/direction.

STEP 5: Multiply the angle from Step 4 by 2 to put this into "Mohr's Circle World" and then construct a new circle.

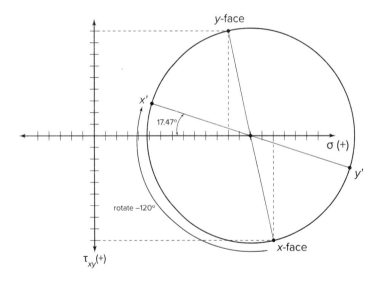

To find out the angle on Mohr's for calculating *x'*, we first need to figure out the angle between the original *x*-face and the positive horizontal axis.

$$tan(\theta) = \frac{45}{10} \qquad \theta = 77.47°$$

Therefore the original *x*-face is 77.47° clockwise from the positive horizontal axis. Then we need to rotate from this *x*-face position 120° clockwise.

−77.47° − 120° = −197.47° so that would mean the *x'* point is located at 197.47° − 180° = 17.47°.

STEP 6: Calculate the new coordinates for the x'-face.

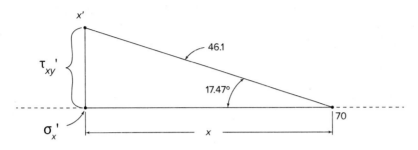

$$cos(17.47°) = \frac{x}{46.1} \Rightarrow x = 43.97$$

$$sin(17.47°) = \frac{\tau_{xy}{}'}{46.1} \Rightarrow \tau_{xy}{}' = -13.84 \text{ MPa}$$

$$\sigma_x{}' = 70 - 43.97 = 26.03 \text{ MPa}$$

STEP 7: Draw the results on a sectional element. The following images all represent the same answer.

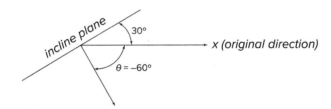

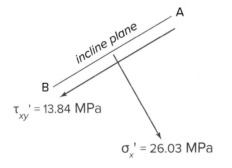

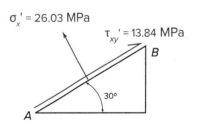

PRO TIP

Although these plane stress problems can be solved using the equation method, Mohr's Circle gives you a much better visual representation and physical feel.

▶ Press pause on video lesson 51 once you get to the workout problem. Only press play if you get stuck.

Find the stresses on plane *AB*.

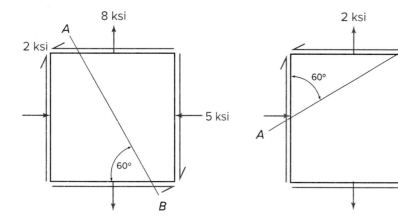

✓ TEST YOURSELF 9.4

For the stress state shown in the figure, compute the normal and shear stress on the face rotated to the plane indicated.

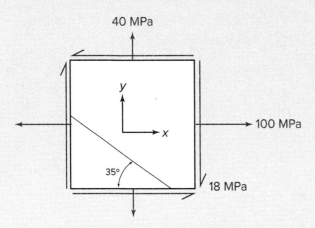

40 MPa

100 MPa

35°

18 MPa

ANSWERS

$\sigma_x' = 42.8$ MPa

$\tau_{xy}' = -22.0$ MPa

> ▶ **WATCH VIDEO** **MECHANICS OF MATERIALS LESSON 52**
> **Deriving Stress Element for Mohr's Circle, Combined Loading**

The reality of being an engineer is that we will not be given the stress element for analysis. As a matter of our job, we are supposed to find the stresses. This will require us to use everything we have learned to derive a stress element at a particular point on a rigid body. After we have that stress state, we can calculate the principal stress, max shear stress, and stress on an inclined plane.

EXAMPLE: STRESS ELEMENT FOR MOHR'S CIRCLE

Determine the principal stress at point *A* located at the top of the web on the loaded beam given. Show results on a stress element.

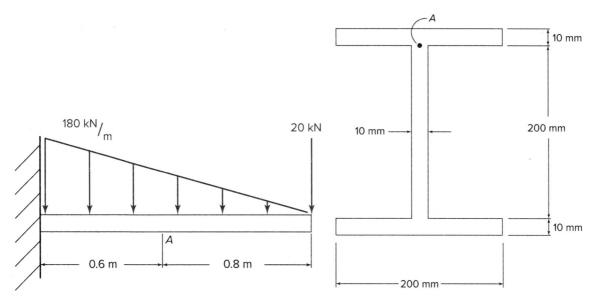

STEP 1: Section the beam through point *A* and draw the FBD of the right-hand side (the right-hand side was chosen so that you don't need to solve for the wall reactions first).

To find the value of the distributed load at point *A*, use similar triangles.

$$\frac{180}{1.4 \text{ m}} = \frac{y}{0.8 \text{ m}}$$

$$y = 102.86 \frac{\text{kN}}{\text{m}}$$

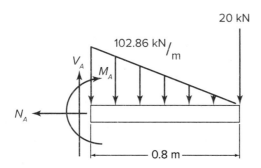

STEP 2: Solve for internal forces M, N, and V.

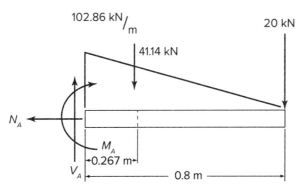

Total load $= \dfrac{1}{2}(0.8)(102.86) = 41.14$ kN

$\Sigma F_x = 0 = N_A, \therefore N_A = 0$

$\Sigma F_y = 0 = V_A - 20 - 41.14 \Rightarrow V_A = 61.14$ kN

$\Sigma M_A = 0 = -M_A - 41.14\left(\dfrac{0.8}{3}\right) - 20(0.8)$

(Don't forget that the centroid for a triangle is ⅓ times the base.)

$M_A = -26.97$ kN•m

Recall the direction of M_A was from the positive sign convention and was simply an assumption of direction. The fact that we got a negative here tells us that our assumption was actually backward.

STEP 3: Now that we have bending moment (M_A) and shear force (V_A) at point A, we can calculate the normal stresses (σ) and shear stress (τ) at point A.

Normal Stress

$\sigma = \dfrac{Mc}{I}$

$I = \dfrac{1}{12}(200)(220)^3 - \dfrac{1}{12}(190)(200)^3 = 50.8 \times 10^6$ mm^4

$c = 100$ mm

$M = 26.97$ kN•m

With the given bending and moment of point A, the top of the beam is in tension and the bottom in compression. So point A will have a positive normal stress.

$\sigma = \dfrac{(26.97 \text{ kN•m})\left(\dfrac{1,000 \text{ N}}{\text{kN}}\right)\left(\dfrac{1,000 \text{ mm}}{\text{m}}\right)(100 \text{ mm})}{50.8 \times 10^6 \text{ mm}^4} = 53.1 \text{ MPa}$

Shear Stress

$$\tau = \frac{VQ}{It}$$

$V = 61.14 \text{ kN}$

$I = 50.8 \times 10^6 \text{ mm}^4$

Area above the web for calculating Q and t:

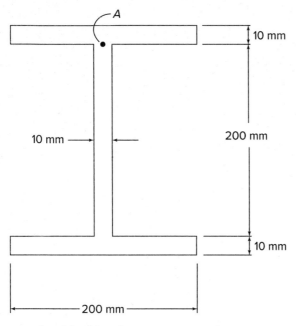

$Q = (105)(10)(200) = 210{,}000 \text{ mm}^3$

$t = 10 \text{ mm}$

$$\tau = \frac{61.14 \text{ kN}(\dfrac{1{,}000 \text{ N}}{\text{kN}})(210{,}000 \text{ mm}^3)}{50.8 \times 10^6 \text{ mm}^4 \,(10 \text{ mm})} = 25.27 \text{ MPa}$$

Shear stress direction image:

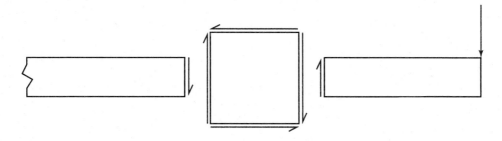

STEP 4: Create the stress element for point A.

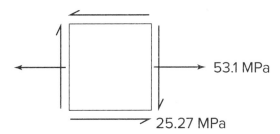

STEP 5: Determine Mohr's Circle coordinate and construct Mohr's Circle.

x-face $(\sigma_x, \tau_{xy}) = x$-face $(53.1, -25.7)$

y-face $(\sigma_y, -\tau_{xy}) = y$-face $(0, 25.27)$

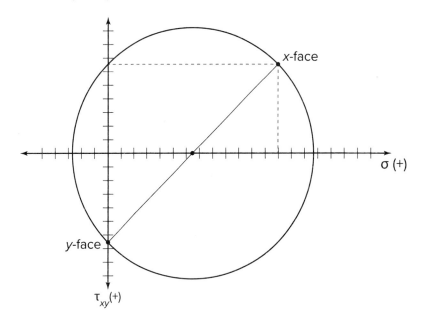

STEP 6: Calculate the center of the circle (σ_{AVG}) and the radius $(\tau_{xy\,max})$.

$$\sigma_{AVG} = \left(\frac{53.1 - 0}{2}\right) = 26.55 \text{ MPa}$$

$$\tau_{xy\,max} = \sqrt{(26.55^2 + 25.27^2)} = 36.65 \text{ MPa}$$

STEP 7: Find principal stresses.

$\sigma_1, \sigma_2 = 26.55 \pm 36.65$

$\therefore \sigma_1 = 63.2 \text{ MPa}, \sigma_2 = -10.1 \text{ MPa}$

Press pause on video lesson 52 once you get to the workout problem. Only press play if you get stuck.

Determine the principal stresses and the absolute max shear stress at A.

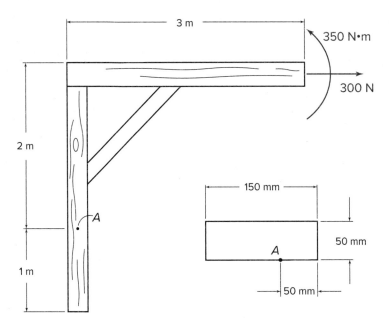

TEST YOURSELF 9.5

SOLUTION
TO TEST
YOURSELF:
Stress
Transformation

For the cantilever beam shown, compute the maximum shear stress 8 ft from the left support at a point 1 inch up from the neutral axis of the cross section as shown.

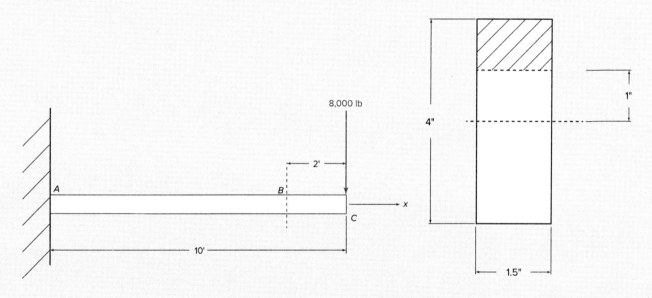

STRESS TRANSFORMATION

WATCH VIDEO   **MECHANICS OF MATERIALS LESSON 53**
Mohr's Circle on Thin-Walled Pressure Vessel

 Press pause on video lesson 53 once you get to the workout problem. Only press play if you get stuck.

Find the principal stresses and max shear stress at *A* for the following pressure vessel if the inner diameter is 0.5 inches and the wall thickness is 0.025 inches.

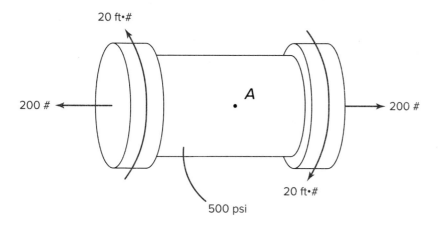

TEST YOURSELF 9.6

SOLUTION TO TEST YOURSELF: Stress Transformation

A segment of a sealed thin-walled tube is pressurized to 3.5 MPa while being pulled with an axial force of 1,000 N and a torque of 30 N·m as shown. The tube has an outside diameter of 30 mm and a wall thickness of 1 mm. What is the maximum in-plane shear stress?

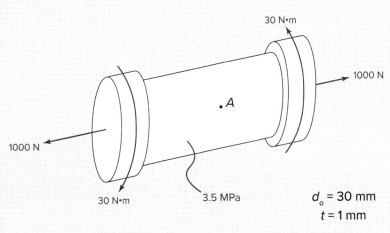

d_o = 30 mm

t = 1 mm

STRESS TRANSFORMATION

ANSWER

$\tau_{xy\ max}$ = 24.5 MPa

MECHANICS OF MATERIALS LESSON 54
Absolute Max Shear Stress with Volume Element

Absolute Max Shear Stress

Up to this point, we have been working with 2D stress elements such as the one given below:

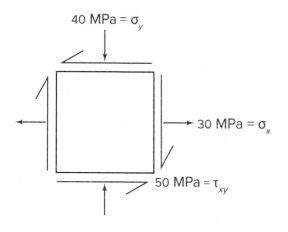

From here, we could construct Mohr's Circle and calculate the principal stresses as well as the "in-plane" maximum shear stress.

Note: When you see the term "in-plane," you are looking for stresses (shear and normal) on that 2D element. *But* is "in-plane" max shear stress the "absolute" max shear stress? The answer is maybe, but maybe not.

There is actually no such thing as a 2D stress element, or small piece of loaded material.

The 2D stress element given above actually looks like this . . . in 3D!

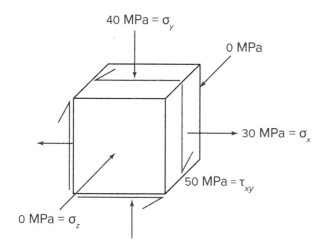

To find "absolute" max shear stress, construct Mohr's Circle (as usual) for the 2D stress element, find principal stresses and "in-plane" max shear stress.

Next construct two more Mohr's Circles for the other two views of the volume element cube (one for side view *xz*, and one for side view *yz*). Although not drawn on this image, τ_{xz} and τ_{yz} are both zero.

Finally, analyze all three views to find which view (*xy*, *yz*, or *xz*) contain the absolute maximum shear stress.

EXAMPLE: ABSOLUTE MAX SHEAR STRESS

Find the absolute maximum shear stress for the following stress element.

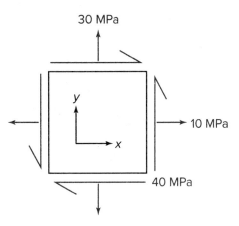

STEP 1: Find the coordinates of Mohr's Circle and construct the circle.

x-face $(\sigma_x, \tau_{xy}) = x$-face $(10, 40)$

y-face $(\sigma_y, -\tau_{xy}) = y$-face $(30, -40)$

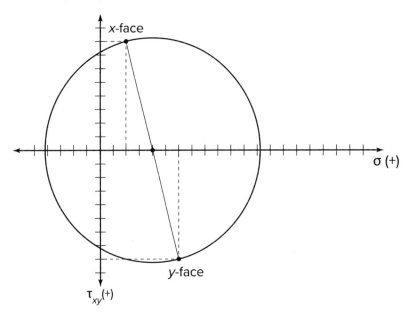

STEP 2: Find the center of the circle (σ_{AVG}) and the radius of the circle ($\tau_{xy\,max}$).

Center of circle = $\sigma_{\text{AVG}} = 10 + \dfrac{30 - 10}{2} = 20$ MPa

Radius of circle = $\tau_{xy\,max} = \sqrt{(10^2 + 40^2} = 41.23$ MPa

STEP 3: Calculate the principal stresses.

$\sigma_1, \sigma_2 = 20 \pm 41.23$

$\sigma_1 = 61.23$ MPa, $\sigma_2 = -21.23$ MPa

STEP 4: Since there is no shear stress on either the *xz*-face or the *yz*-face, the normal stresses on those faces will be the principal stresses.

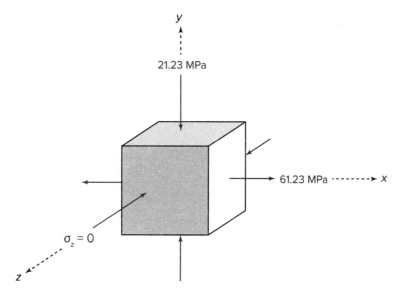

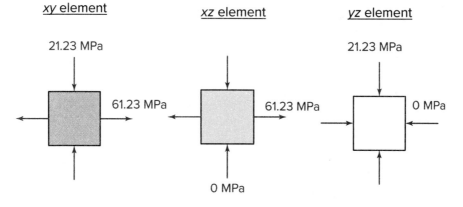

STEP 5: Plot all three circles on the same graph.

	σ_1	σ_2
Circle 1	(61.23,	−21.23)
Circle 2	(61.23,	0)
Circle 3	(−21.23,	0)

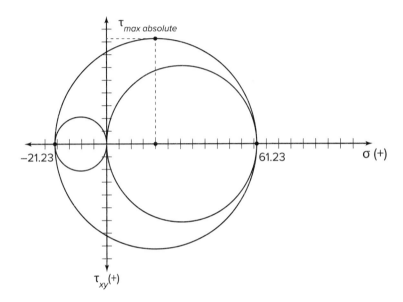

Note: If Mohr's Circle straddles the τ axis (first coordinate of σ in $+\sigma$ territory while the second coordinate is in $-\sigma$ territory), then the "in-plane" max shear stress will be the "absolute" max shear stress.

STEP 6: As shown in the graph above, the first circle (always the biggest circle!) contains the "absolute max shear stress," in this case, the radius of the biggest circle.

$$\tau_{absolute\ max} = 41.23 \text{ MPa}$$

Press pause on video lesson 54 once you get to the workout problem. Only press play if you get stuck.

Find the absolute max shear stress for the following stress element.

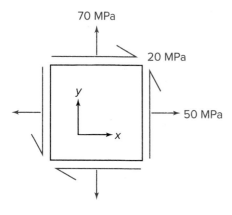

TEST YOURSELF 9.7

SOLUTION TO TEST YOURSELF: Stress Transformation

For the stress state shown in the figure, what is the absolute maximum shear stress? What are the three principal stresses?

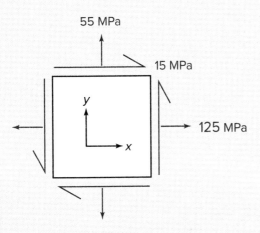

STRESS TRANSFORMATION

ANSWERS

$\tau_{absolute\ max} = 64$ MPa

$\sigma_1 = 128.1$ MPa

$\sigma_2 = 51.9$ MPa

$\sigma_3 = 0$

MECHANICS OF MATERIALS LESSON 55
Tresca, Von Mises, and Rankine Failure Theories Explained

We will now explore ways we can identify failure of a material under complex loading, in our case, we will consider 2D plane stress (that is, we only have σ_x, σ_y, τ_{xy}).

When we talk about materials failing under load, we commonly think of two distinct situations. The first is a *ductile failure* and the second is a *brittle failure*.

Ductile Failure

For ductile failure, we base "failure" on yielding of the material as determined from a 1D tensile test (you can refresh on this in Level 3). However, a ductile failure does not necessarily mean that the part breaks into two pieces! It fails by yielding according to a *failure theory*. We will consider two failure theories for ductile materials: **Tresca** and **von Mises**.

- **Tresca Failure Theory.** This theory is based on the *critical shear stress* approach. It is sometimes called the *maximum shear stress failure theory*. As this implies, the theory states that failure occurs when the maximum shear stress in a material reaches the shear stress at yielding in a 1D tension or compression test (sometimes a simple shear test is used). When we write this in terms of the principal stresses, the Tresca Failure Theory states that no yielding (failure) will occur if the following conditions are met where σ_1, σ_2 are the principal stresses and the yield stress is given as σ_y.

$$\left.\begin{array}{l} |\sigma_1| < \sigma_y \\ |\sigma_2| < \sigma_y \end{array}\right\} \text{ if } \sigma_1 \text{ and } \sigma_2 \text{ have the same signs}$$

$$|\sigma_1 - \sigma_2| < \sigma_y \} \text{ if } \sigma_1 \text{ and } \sigma_2 \text{ have opposite signs}$$

- **von Mises Failure Theory.** This theory is based on the energy (work) that causes *distortion* (that is, not simply shrinkage or expansion). This theory states that failure occurs when the distortion energy per unit volume exceeds the distortion energy at yielding in a 1D test. As with Tresca, both these theories benchmark failure to a 1D material test (typically a tensile test). The von Mises Failure Theory is sometimes called the *distortional strain energy theory* and states that no yielding (failure) will occur if the following condition is met:

$$\sqrt{\sigma_1^2 + \sigma_2^2 - (\sigma_1)(\sigma_2)} < \sigma_y$$

Brittle Failure

For *brittle failure* (materials that don't display much plastic behavior after yielding, such as glass and concrete), we base failure on rupture and therefore use the ultimate stress σ_{ult}.

- **Rankine Failure Theory (Maximum Normal Stress Theory).** This theory suggests that failure occurs when the maximum principal stress in a material reaches the *ultimate strength* of the material in tension or compression. This can be expressed as:

$$|\sigma_1| < \sigma_{ult}$$
$$|\sigma_2| < \sigma_{ult}$$

PRO TIP

- For ductile materials, use Tresca or von Mises.
- For brittle materials, use Rankine.
- Take care in computing the principal stresses as they are the key ingredient to the failure theories.

EXAMPLE: FAILURE THEORIES

For the given stress state in a ductile material, determine if there is failure according to Tresca and von Mises. The yield stress from a 1D tensile test is 82 MPa.

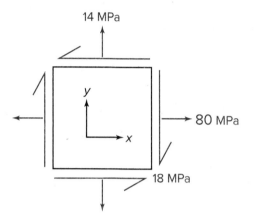

14 MPa

80 MPa

18 MPa

STEP 1:   Compute the principal stresses.

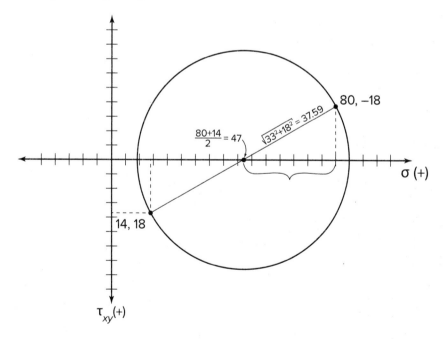

STEP 2: Investigate Tresca and von Mises failure conditions:

$$\sigma_1 = 47 + 37.59 = 84.6 \text{ MPa}$$

$$\sigma_2 = 47 - 37.59 = 9.41 \text{ MPa}$$

Tresca:

$$|\sigma_1| = |84.6 \text{ MPa}| > 82 \text{MPa}$$

$\therefore$ yield predicted

von Mises:

$$\sqrt{(84.6 \text{ MPa})^2 + (9.41 \text{ MPa})^2 - (84.6 \text{ MPa})(9.41 \text{ MPa})}$$

$$= 80.3 \text{ MPa} < 82 \text{ MPa}$$

$\therefore$ no yielding

STEP 3: Review the solution. As we can see, the two failure theories do not agree. Generally, for most ductile materials, von Mises is more accurate at predicting yield while Tresca is more conservative.

Press pause on video lesson 55 once you get to the workout problem. Only press play if you get stuck.

1. Find the smallest radius in the cast iron shaft so that it does not fail due to max normal stress theory.

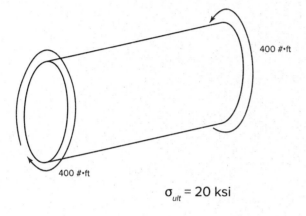

400 #·ft

400 #·ft

σ_{ult} = 20 ksi

2. Will the following specimen fail due to the max shear stress theory and/or the max distortion energy theory?

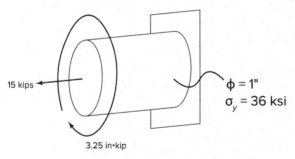

15 kips

ϕ = 1"
σ_y = 36 ksi

3.25 in·kip

TEST YOURSELF 9.8

SOLUTION TO TEST YOURSELF: Stress Transformation

A thin-walled cylindrical vessel made from a ductile material is pressurized to 5 MPa with an outer diameter of 420 mm and a wall thickness of 10 mm. If the yield stress is 95 MPa, check to see if it will fail by Tresca and von Mises.

STRESS TRANSFORMATION

ANSWERS
Tresca: yes
von Mises: no

In this chapter we explored how to find the stress on an element at any desired angle.

- Two techniques for solving: (1) equation method and (2) Mohr's circle.
- Two common states to solve for:
  - Principal stress—when the normal stress is at a maximum/minimum and shear is zero
  - Maximum shear stress—occurs when shear stress is maximum and sigma stress is at the average stress

    Note: You need to be careful when solving for the maximum shear stress as it may actually occur out-of-plane.
  - Once you have solved for the principal stresses, you can then try to determine whether failure occurs or not via the failure theories (Tresca, von Mises, and Rankine).

Stress Transformation Equations

$$\sigma_x' = \frac{\sigma_x + \sigma_y}{2} + \frac{\sigma_x - \sigma_y}{2} \cos(2\theta) + \tau_{xy} \sin(2\theta)$$

$$\sigma_y' = \frac{\sigma_x + \sigma_y}{2} - \frac{\sigma_x - \sigma_y}{2} \cos(2\theta) - \tau_{xy} \sin(2\theta)$$

$$\tau_{xy}' = -(\frac{\sigma_x - \sigma_y}{2}) \sin(2\theta) + \tau_{xy} \cos(2\theta)$$

Max Shear and Principal Stresses

$$\sigma_1, \sigma_2 = \frac{\sigma_x + \sigma_y}{2} \pm \sqrt{\left(\frac{\sigma_x - \sigma_y}{2}\right)^2 + \tau_{xy}^2}$$

$$\tau_{xy\,max} = \pm \sqrt{\left(\frac{\sigma_x - \sigma_y}{2}\right)^2 + \tau_{xy}^2}$$

Equations for Principal Stress Angles (Θp) as well as Maximum Shear Stress Angles (Θs)

$$\tan(2\theta_p) = \frac{\tau_{xy}}{\left(\frac{\sigma_x - \sigma_y}{2}\right)}$$

$$\tan(2\theta_s) = -\left[\frac{\left(\frac{\sigma_x - \sigma_y}{2}\right)}{\tau_{xy}}\right]$$

Failure Theories

Tresca (Maximum Shear Stress Theory)

$$|\sigma_1| < \sigma_y$$
$$|\sigma_2| < \sigma_y$$ $\Big\}$ if σ_1 and σ_2 have the same signs

$$|\sigma_1 - \sigma_2| < \sigma_y \ \} \ \text{if } \sigma_1 \text{ and } \sigma_2 \text{ have opposite signs}$$

von Mises Failure Theory

$$\sqrt{\sigma_1^2 + \sigma_2^2 - (\sigma_1)(\sigma_2)} < \sigma_y$$

Rankine Failure Theory

$$|\sigma_1| < \sigma_{ult}$$
$$|\sigma_2| < \sigma_{ult}$$

PRO TIPS

Stress Transformation Equations

- Make sure to pay attention to the sign on the angle when using the stress transformation equations. If the angle is clockwise as measured from the x to x' axis, then the angle is negative. If the angle is counterclockwise as measured from the x to x' axis, then the angle is positive.

- When using the equations to calculate stress transformations, be super careful with your signs as this will change your answer. Calculating using Mohr's Circle will give you a better visualization of your answers and reduce mistakes.

- The magnitude of the principal stress angles (θ_{p_1} and θ_{p_2}) will always add to 90°.

    Likewise, the magnitude of the maximum shear stress angles (θ_{s_1} and θ_{s_2}) will also always add to 90°.

Mohr's Circle

- To create Mohr's circle, the key is simply to flip the sign on shear for one of the points that you plot (it doesn't matter which one). This allows you to create a circle that is centered on the x-axis.

    Then, make sure you pay attention to what shear you paired with your σ_x and σ_y coordinate axes. Different instructors use different signs conventions, so be consistent (especially pay attention to when you are rotating your stress element).

- A straightedge and a circle drawing tool such as a compass are a must for generating good drawings and enhancing your understanding.

- Although these plane stress problems can be solved using the equation method, Mohr's Circle gives you a much better visual representation and physical feel.

Failure Theories

- For ductile materials, use Tresca or von Mises.
- For brittle materials, use Rankine.
- Take care in computing the principal stresses as they are the key ingredient to the failure theories.

PITFALLS

Mohr's Circle
- Your final transformation angle should *always* be reported in "real world" degrees, not in "Mohr's Circle world" degrees.

Level 10
Strain Transformation

Strain Transformations

Strain Rosettes

STRAIN TRANSFORMATION

 WATCH VIDEO | **MECHANICS OF MATERIALS LESSON 56**
Strain Transformation with Equations and Mohr's Circle

Strain Transformations

Once you have mastered stress transformations, strain transformations should be pretty straightforward to master as well. Strain transformations can be completed with either the equation method or Mohr's Circle with only slight modifications.

Equation Method

$$\epsilon_x' = \frac{\epsilon_x + \epsilon_y}{2} + \frac{\epsilon_x - \epsilon_y}{2} cos(2\theta) + \frac{\gamma_{xy}}{2} sin(2\theta)$$

$$\epsilon_y' = \frac{\epsilon_x + \epsilon_y}{2} - \frac{\epsilon_x - \epsilon_y}{2} cos(2\theta) - \frac{\gamma_{xy}}{2} sin(2\theta)$$

$$\frac{\gamma_{xy}'}{2} = -(\frac{\epsilon_x - \epsilon_y}{2}) sin(2\theta) + \frac{\gamma_{xy}}{2} cos(2\theta)$$

$$\epsilon_1, \epsilon_2 = \frac{\epsilon_x + \epsilon_y}{2} \pm \sqrt{(\frac{\epsilon_x - \epsilon_y}{2})^2 + (\frac{\gamma_{xy}}{2})^2}$$

$$tan(2\theta_p) = \frac{\gamma_{xy}}{(\epsilon_x - \epsilon_y)}$$

$$tan(2\theta_s) = -\frac{(\epsilon_x - \epsilon_y)}{\gamma_{xy}}$$

These equations should look very similar to the stress transformation equations. Everywhere we had a σ, we now have an ϵ, and everywhere we had a τ_{xy} we now have $\frac{\gamma_{xy}}{2}$.

We will not be working examples using the equation method in this text. If you need to review these techniques, review Lessons 48 and 49 and simply make the following substitutions:

$$\sigma_x = \epsilon_x \qquad \sigma_y = \epsilon_y \qquad \tau_{xy} = \frac{\gamma_{xy}}{2}$$

PRO TIP

Similar to before, make sure you remember how θ is defined in these equations. It is measured from the x-axis to the x' axis.

Strain Transformations Using Mohr's Circle

Just like the equation method for strain transformation, Mohr's circle works essentially the same for strain as it did for stress. We simply need to make some substitutions.

Where we had σ_x and σ_y, we now substitute ϵ_x and ϵ_y. And where we had τ_{xy} for stress, we have $\frac{\gamma_{xy}}{2}$ for strain. Don't forget that the shear strain angle is divided by 2 for Mohr's circle. Notice the similarities in the method of Mohr's circle in the following example.

EXAMPLE: STRAIN TRANSFORMATION

If the state of plane strain at a particular point has components of $\epsilon_x = -350\,\mu$, $\epsilon_y = -110\,\mu$, and $\gamma_{xy} = 120\,\mu$. Determine the state of strain on an element oriented at 25° counterclockwise from this position.

STEP 1:   Determine the coordinates to plot Mohr's Circle.

$$x\text{-face}\ (\epsilon_x, \frac{\gamma_{xy}}{2}) = (-350, \frac{120}{2}) = (-350, 60)$$

$$y\text{-face}\ (\epsilon_x, \frac{-\gamma_{xy}}{2}) = (-110, \frac{-120}{2}) = (-110, -60)$$

STEP 2:   Plot the values in "Mohr's space." Be sure to remember that positive is in the downward direction (at least that is how we are going to teach it in this book)!

Note: Also calculate the center and the radius of the circle.

$$r = \sqrt{(120^2 + 60^2)} = 134.16$$

$$C = \frac{350 - 110}{2} = 120 \qquad -110 - 120 = -230\ (\text{center point})$$

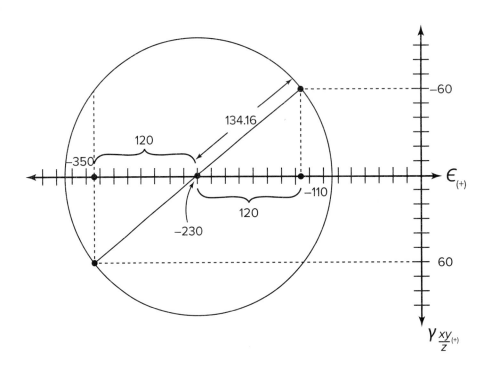

STEP 3: Rotate the points 50° clockwise. (***Recall:*** To rotate angles in "Mohr's Circle World," we must multiply the "real-world" angle [25°] by 2.)

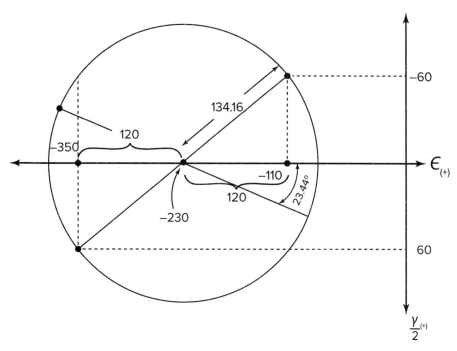

$$tan(\theta) = \frac{60}{120}$$

$$\theta = 26.56°$$

Rotate clockwise 50° which will rotate it to 50° − 26.56° = 23.44° below the ϵ-axis

STEP 4: Calculate the new values of strain.

134.16 * cos(23.44°) = 123.09

134.16 * sin(23.44°) = 53.37

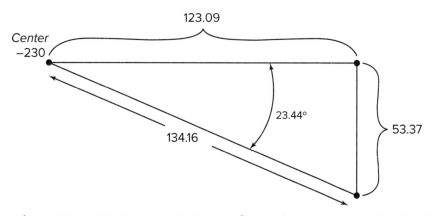

$\epsilon_x' = -230 + 123.09 = -106.91$ μ   (Read this as negative 106.91 microstrains.)

$\epsilon_y' = -230 - 123.09 = -353.09$ μ

$\gamma_{xy} = 53.37 * 2 = 106.74$ μ

▶ Press pause on video lesson 56 once you get to the workout problem. Only press play if you get stuck.

Find the principal strains and max shear strain.

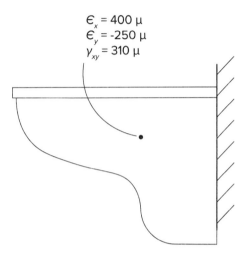

$\epsilon_x = 400\ \mu$
$\epsilon_y = -250\ \mu$
$\gamma_{xy} = 310\ \mu$

✓ **TEST YOURSELF 10.1**

SOLUTION TO TEST YOURSELF: Strain Transformation

At the point in the part as shown, compute the principal strains and the maximum in-plane shear strain.

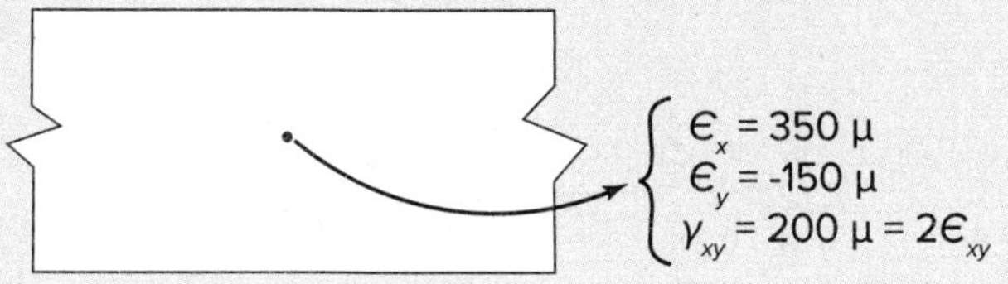

$$\epsilon_x = 350\ \mu$$
$$\epsilon_y = -150\ \mu$$
$$\gamma_{xy} = 200\ \mu = 2\epsilon_{xy}$$

ANSWERS

$$\epsilon_1 = 369.3\ \mu$$
$$\epsilon_2 = -169.3\ \mu$$
$$\gamma_{xy} = 538.6\ \mu$$

STRAIN TRANSFORMATION

 WATCH VIDEO **MECHANICS OF MATERIALS LESSON 57**
Strain Gauge Rosettes Explained

How do we measure strain?

- **Strain**
  - Deformation or a change in length.
  - Measured with a little gauge that tells how much something is stretched.
- **Strain Gauge**
  - Extremely tiny gauge (smaller than your fingertip) made out of mylar film.
  - Mounted with epoxy onto the surface of the object.
  - Contains a wire that runs back and forth as shown in the image.
  - As you stretch the gauge, the wire stretches, which results in a change in resistance.
  - It works in both tension and in compression.
  - Unfortunately it works only in this axial direction. Thus if you want to know more, you will need to mount multiple of these on your sample.
- **Strain Rosettes**
  - Arrangement of a whole bunch of strain gauges at a known angle.
  - The two most common strain rosette configurations are 45° from one another and 120° from one another.
  - Allows you to determine the entire strain state of an element through the following equations:

$$\epsilon_a = \epsilon_x cos^2(\theta_a) + \epsilon_y sin^2(\theta_a) + \gamma_{xy} sin(\theta_a)cos(\theta_a)$$
$$\epsilon_b = \epsilon_x cos^2(\theta_b) + \epsilon_y sin^2(\theta_b) + \gamma_{xy} sin(\theta_b)cos(\theta_b)$$
$$\epsilon_c = \epsilon_x cos^2(\theta_c) + \epsilon_y sin^2(\theta_c) + \gamma_{xy} sin(\theta_c)cos(\theta_c)$$

  - ϵ_a, ϵ_b, and ϵ_c are readings directly from the strain gauges. If you know the angle these were mounted at, then you can find ϵ_x, ϵ_y, and γ_{xy}.
  - Once you know ϵ_x, ϵ_y, and γ_{xy}, then you can perform Mohr's circle to find the stress at any desired orientation.

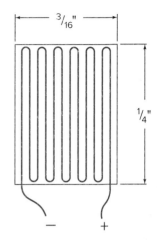

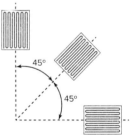

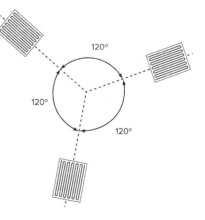

 **MECHANICS OF MATERIALS LESSON 58**
Strain Rosette Example Problem with Mohr's Circle

 # EXAMPLE: STRAIN ROSETTE W/MOHR'S CIRCLE

The given strain rosette was placed on a tractor component to measure strain. The following readings were observed:

$\epsilon_a = 160 \times 10^{-6}$

$\epsilon_b = 100 \times 10^{-6}$

$\epsilon_c = 80 \times 10^{-6}$

Find the principle in-plane strains and the in-plane maximum shear strain at the point of application.

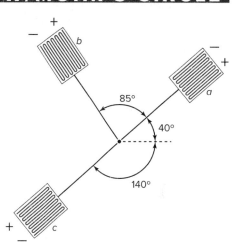

STEP 1: From the strain rosettes, determine the following. Note that all of these angles are always referenced from positive *x*-axis rather than each other.

$\epsilon_a = 160 \times 10^{-6}$ $\theta_a = 40°$

$\epsilon_b = 100 \times 10^{-6}$ $\theta_b = 125°$

$\epsilon_c = 80 \times 10^{-6}$ $\theta_c = (360° - 140°) = 220°$

STEP 2: Substitute into the strain rosette equations and solve using system solver.

$\epsilon_a = \epsilon_x cos^2(\theta_a) + \epsilon_y sin^2(\theta_a) + \gamma_{xy} sin(\theta_a)cos(\theta_a)$

$\epsilon_b = \epsilon_x cos^2(\theta_b) + \epsilon_y sin^2(\theta_b) + \gamma_{xy} sin(\theta_b)cos(\theta_b)$

$\epsilon_c = \epsilon_x cos^2(\theta_c) + \epsilon_y sin^2(\theta_c) + \gamma_{xy} sin(\theta_c)cos(\theta_c)$

$160 \times 10^{-6} = \epsilon_x cos^2(40°) + \epsilon_y sin^2(40°) + \gamma_{xy} sin^2(40°)cos^2(40°)$

$100 \times 10^{-6} = \epsilon_x cos^2(125°) + \epsilon_y sin^2(125°) + \gamma_{xy} sin^2(125°)cos^2(125°)$

$80 \times 10^{-6} = \epsilon_x cos^2(220°) + \epsilon_y sin^2(220°) + \gamma_{xy} sin^2(220°)cos^2(220°)$

Using a System Solver:

$\epsilon_x = 0.816 \times 10^{-6} = 0.816 \mu$

$\epsilon_y = 0.569 \times 10^{-6} = 0.569 \mu$

$\gamma_{xy} = -2.944 \times 10^{-6} = -2.944 \mu$

▶ Press pause on video lesson 58 once you get to the workout problem. Only press play if you get stuck.

D ue to loadings, $\epsilon_a = 60 \times 10^{-6}$, $\epsilon_b = 135 \times 10^{-6}$, $\epsilon_c = 264 \times 10^{-6}$, find the principal strains.

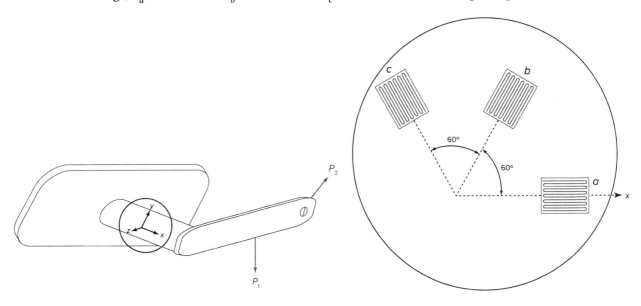

TEST YOURSELF 10.2

SOLUTION TO TEST YOURSELF: Strain Transformation

STRAIN TRANSFORMATION

For the rectangular strain rosette shown, compute the maximum in-plane shear strain.

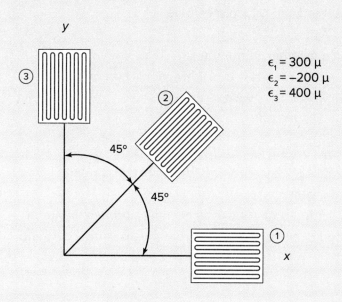

$\epsilon_1 = 300 \ \mu$
$\epsilon_2 = -200 \ \mu$
$\epsilon_3 = 400 \ \mu$

ANSWER

$\gamma_{max} = 1,104.5 \ \mu$

Strain Transformation

In this chapter we explored how to find the strain on an element at any desired angle.

- Two techniques for solving: (1) equation method and (2) Mohr's Circle.
- A common state to solve for is:
  - Principal strain—when the strain is at a maximum/minimum and shear strain is zero
- The strain transformations are nearly the same as the stress transformations. Everywhere we had a σ, we now have an ϵ, and everywhere we had a τ_{xy} we now have $\dfrac{\gamma_{xy}}{2}$.

Strain Rosettes

- Arrangement of a whole bunch of strain gauges at a known angle. There are two common orientations: 45° and 120° between the strain gauges.
- Allows you to determine ϵ_x, ϵ_y, and γ_{xy} (i.e., entire strain state of an element) through the strain rosette equations (i.e., by knowing ϵ_a, ϵ_b, and ϵ_c, the strain output from the measuring device).

Strain Transformation Equations

$$\epsilon_x' = \frac{\epsilon_x + \epsilon_y}{2} + \frac{\epsilon_x - \epsilon_y}{2}\cos(2\theta) + \frac{\gamma_{xy}}{2}\sin(2\theta)$$

$$\epsilon_y' = \frac{\epsilon_x + \epsilon_y}{2} - \frac{\epsilon_x - \epsilon_y}{2}\cos(2\theta) - \frac{\gamma_{xy}}{2}\sin(2\theta)$$

$$\frac{\gamma_{xy}'}{2} = -\left(\frac{\epsilon_x - \epsilon_y}{2}\right)\sin(2\theta) + \frac{\gamma_{xy}}{2}\cos(2\theta)$$

$$\epsilon_1, \epsilon_2 = \frac{\epsilon_x + \epsilon_y}{2} \pm \sqrt{\left(\frac{\epsilon_x - \epsilon_y}{2}\right)^2 + \left(\frac{\gamma_{xy}}{2}\right)^2}$$

$$\tan(2\theta_p) = \frac{\gamma_{xy}}{(\epsilon_x - \epsilon_y)}$$

$$\tan(2\theta_s) = -\frac{(\epsilon_x - \epsilon_y)}{\gamma_{xy}}$$

Strain Rosettes

$$\epsilon_a = \epsilon_x cos^2(\theta_a) + \epsilon_y sin^2(\theta_a) + \gamma_{xy} sin(\theta_a)cos(\theta_a)$$
$$\epsilon_b = \epsilon_x cos^2(\theta_b) + \epsilon_y sin^2(\theta_b) + \gamma_{xy} sin(\theta_b)cos(\theta_b)$$
$$\epsilon_c = \epsilon_x cos^2(\theta_c) + \epsilon_y sin^2(\theta_c) + \gamma_{xy} sin(\theta_c)cos(\theta_c)$$

PRO TIP

Strain Transformation

Similar to before, make sure you remember how θ is defined in these equations. It is measured from the *x*-axis to the *x*' axis.

Level 11
Design of Beams and Shafts

MECHANICS OF MATERIALS LESSON 59
Introduction to Beam Design Example Problem

Beam Design

We will introduce how to design beams for particular applications. Whether you are designing support beams for an open concept kitchen remodel, a high-rise skyscraper's red-iron skeleton, or a piece of equipment designed to support heavy loads, beam design is an important part of an engineer's toolbox.

Lucky for us, at this point in Mechanics of Materials, we have acquired all the tools necessary to design these beams. As we know, beams need to be the correct size to support the needed loads and not exceed the max allowable normal stresses or maximum shear stress.

Tools required for beam design:

- Ability to create shear and moment diagrams (to find M_{max} and V_{max})
- Ability to calculate the second moment area of inertia for the cross section of your beam (to find I)
- Ability to calculate the maximum normal and shear stress due to bending:
  - $\sigma = \dfrac{Mc}{I}$
  - $\tau = \dfrac{VQ}{It}$

PRO TIP

It is usually desired to find the maximum compressive stress and maximum tensile stress in a beam. Some materials behave differently in tension than in compression (i.e., concrete, which is terrible in tension but good in compression).

EXAMPLE: BEAM DESIGN

The wooden I-beam is supporting the given load. Knowing the allowable bending stress is 12 MPa and the allowable shear stress is 0.8 MPa, find the width (*w*) of the beam so it can safely support the load.

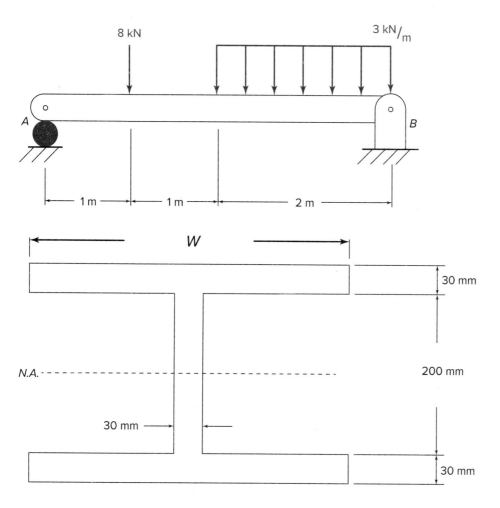

STEP 1: Find global equilibrium.

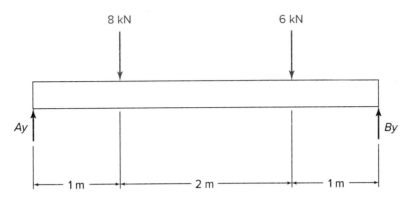

$\Sigma M_A = 0 = B_y(4) - 8(1) - 6(3)$

$B_y = 6.5 \text{ kN}$

$\Sigma F_y = 0 = A_y + 6.5 - 8 - 6 = 0$

$A_y = 7.5 \text{ kN}$

STEP 2: Create a shear-moment diagram for the beam and determine the max shear force (V) and the max bending moment.

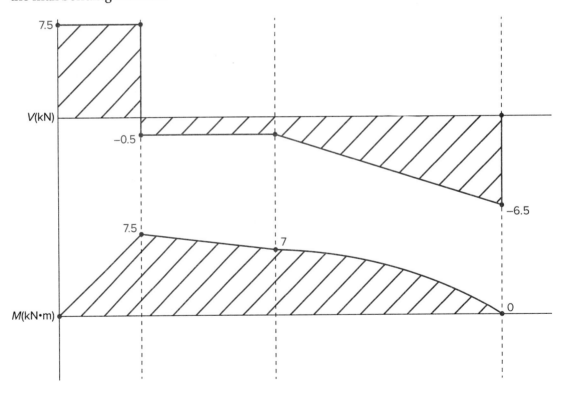

$V_{max} = 7.5 \text{ kN}$

$M_{max} = 7.5 \text{ kN·m}$

STEP 3: Calculate the neutral axis of the beam and the worst-case c.

Via inspection:

$\bar{y} = (200 + 30 + 30)/2 = 130$

$c = \bar{y} = 130$ mm

STEP 4: Calculate I.

$I = \dfrac{1}{12}(w)(260)^3 - \dfrac{1}{12}(w - 30)(200)^3$

$I = 1{,}464{,}666.7w - 666{,}667(w - 30)$

$I = 798{,}000w - 20{,}000{,}010$

STEP 5: Use max allowable bending stress ($\sigma_{allowable} = 12$ MPa) to determine w using the flexure formula.

$$12\text{ MPa} = \frac{(7.5\text{ kN•m})(\dfrac{1{,}000\text{ N}}{\text{kN}})(\dfrac{1{,}000\text{ mm}}{\text{m}})(130\text{ mm})}{798{,}000w - 20{,}000{,}010}$$

$9{,}576{,}000w - 240{,}000{,}120 = 975{,}000{,}000$

$9{,}576{,}000w = 1{,}215{,}000{,}120$

$w = 126.9$ mm

STEP 6: Ensure this beam's shear stress is less than the max allowable shear stress.

$Q = (50)(30)(100) + (115)(30)(126.9) = 587{,}805$ mm³

$t = 30$ mm

$\tau = \dfrac{VQ}{It} = \dfrac{(7{,}500\text{ N})(587{,}805\text{ mm}^3)}{((798{,}000)(126.9) - 20{,}000{,}010\text{ mm}^4)(30\text{ mm})} = 1.81$ MPa so it survives!

▶ Press pause on video lesson 59 once you get to the workout problem. Only press play if you get stuck.

A simply supported timber beam has $\sigma_{allowable}$ = 6.5 MPa and $\tau_{allowable}$ = 500 kPa. Find the dimensions of the beam if it is rectangular having a height to width ratio of 1.25.

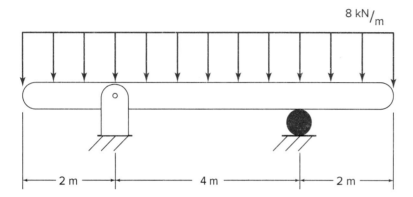

✓ TEST YOURSELF 11.1

SOLUTION TO TEST YOURSELF: Design of Beams and Shafts

For the rectangular cross section loaded as shown, what is the depth of the beam, d, to the nearest ½" if the allowable normal stress is 50 ksi and the allowable shear stress is 29 ksi? Which mode governs the design: bending normal stress or bending shear stress?

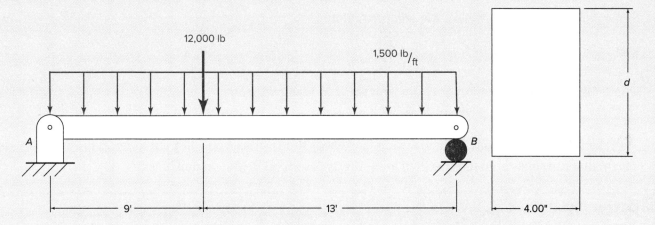

DESIGN OF BEAMS AND SHAFTS

Bending normal stress

$d = 7.5$ in

ANSWERS

MECHANICS OF MATERIALS LESSON 60
Beam Design Section Modulus Tips and Tricks

Section Modulus

The section modulus is a geometric property of the cross section of a beam which we use to select a beam for a given loading application. We can use things such as the wide flange beam tables and use a parameter S in the lookup tables. Recall that $\sigma = \dfrac{Mc}{I}$ and for beam design, we lump the geometric terms into one unknown called S the section modulus.

$$S = \frac{M_{max}}{\sigma_{allowable}} \quad \text{or} \quad S = \frac{I}{c}$$

where M is the maximum moment from your shear moment diagram

$\quad \sigma_{allowable}$ is typically the yield stress with a FoS

When you select a beam, you are paying for steel by the lb. Thus, you will want to select the lightest weight that can safely hold the specified load. This is described in more detail in lesson 61.

PRO TIP	When you determine **S**, don't forget that it is in the denominator of the bending stress equation. Thus, you will want to pick an **S** on the lookup tables that is equal to or larger than the one calculated.

Shear Stress Shortcuts

When we are designing a beam, one of the constraints will be allowable shear stress. Two of the most common beams are:

1. Rectangular cross-sections
2. I-beams (wide flange beams)

Rectangular Shortcut

- Often in design you will use framing lumber, with nominal dimensions such as 2×4 and 2×6 (i.e., an actual 2×4 at the hardware store is 1.5×3.5).
- When calculating Q, you will notice that maximum shear stress happens at the centroid or neutral axis, and when determining this value you can use the approximation that:

$$\tau = 1.5 \left(\frac{V}{A}\right)$$

I-Beam Shortcut

- Used often in larger scale building projects
- When calculating Q, you will notice that maximum shear stress happens at the centroid or neutral axis, and when determining this value you can use the approximation that:

$\tau = (\dfrac{V}{A})$ where A is the web's depth times its thickness (i.e., $A = (d)(t)$) (these dimensions can be found in the table)

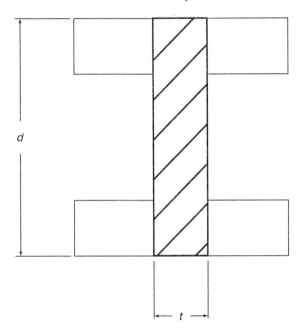

PRO TIP

Note when you purchase lumber, you are purchasing dressed lumber, which has ¼" faced off each side so it's not actually 2 x 4 but rather 1.5 inches x 3.5 inches.

EXAMPLE: RECTANGULAR BEAM SHORTCUT

Determine the minimum dimension b to the nearest mm to safely support the load. The allowable axial stress is 12 MPa, and the allowable shear stress is 1.5 MPa.

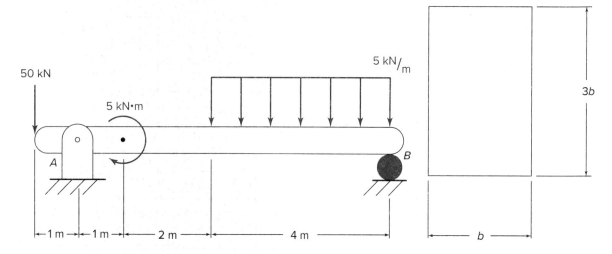

STEP 1: Find global equilibrium.

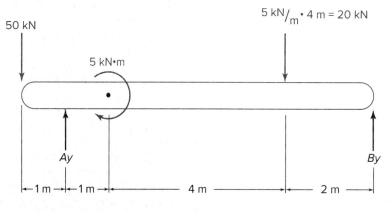

$$\Sigma M_A = 0 = 50(1) - 5 - 20(5) + B_y(7)$$
$$B_y = 7.86 \text{ kN}$$

$$\Sigma F_y = 0 = 7.86 - 20 - 50 + A_y$$
$$A_y = 62.14 \text{ kN}$$

STEP 2: Compute the shear-moment diagram for the loaded beam. Find the max shear force (V_{max}) and the max bending moment (M_{max}).

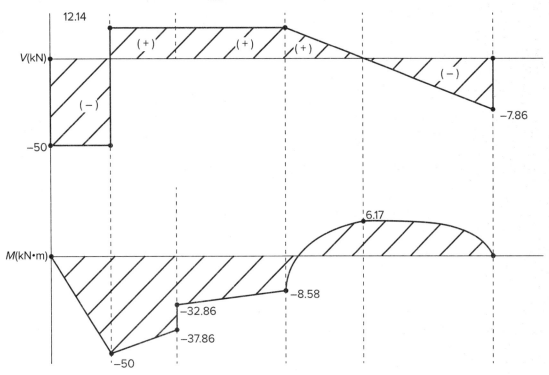

$V_{max} = 50$ kN

$M_{max} = 50$ kN•m

STEP 3: Calculate I for the beam.

$I = (\frac{1}{12}) \times b \times (3b)^3$

$I = 2.25 \times b^4$

STEP 4: Solve for b using the flexural formula ($\sigma = \dfrac{Mc}{I}$) and the allowable normal stress ($\sigma_{allowable} = 12$ MPa).

$$12 \text{ MPa} = \frac{(50 \text{ kN•m})(\dfrac{1{,}000 \text{ mm}}{\text{m}})(\dfrac{3b}{2})}{2.25b^4}$$

$27b^4 = 75{,}000{,}000b$

$b^3 = 2{,}777{,}777.8$

$b = 140.57$ mm

STEP 5: Check results against the allowable shear stress using the transverse shear stress formula ($\tau = \dfrac{VQ}{It}$) and the allowable shear stress ($\tau_{allowable} = 1.5$ MPa).

Here we will use our shortcut method for rectangular beams in which

$$\tau = \frac{VQ}{It} = 1.5\frac{V}{A}$$

$$1.5\,\text{MPa} = 1.5\,\frac{(50\,\text{kN})(\dfrac{1{,}000\,\text{N}}{\text{kN}})}{(b)(3b)}$$

$$3b^2 = 50{,}000$$
$$b^2 = 16{,}666.7$$
$$b = 129.1\,\text{mm}$$

STEP 6: Analyze the final results.

In order to survive $\sigma_{criteria}$, $b \geq 140.57$ mm

In order to survive $\tau_{criteria}$, $b \geq 129.1$ mm

We select the larger value and round up to the next whole mm.

$b = 141$ mm

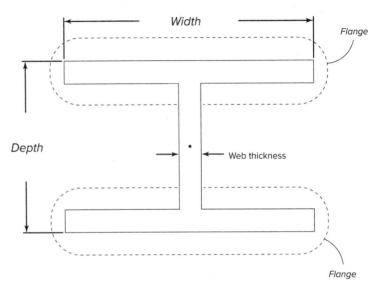

WATCH VIDEO
MECHANICS OF MATERIALS LESSON 61
Wide Flange Beam Design Section Modulus

The designation for wide flange beams is W # x #. The first # tells you nominal depth. The last # is the weight per unit length of the cross section in lb/ft or N/m depending on what table you are using (English or Metric). This is important as it is essentially determining the weight of the beam.

Shown on the right is an idealized sketch of these dimensions, and values for typical wide flange I-beams are shown in the table on the next page.

When looking at these tables, the depth d is the distance from the top of the beam to the bottom. The "web" is the region that connects the two end "flanges." For the flanges, t_f is the thickness and b_f is the width. The web thickness is t_w.

It is also important to note that an I-beam is called an I-beam and not an H-beam because the goal is to maximize your second moment of area (moment of inertia) perpendicular to the load, which is largest when in the "I" orientation. **Therefore, for all situations where we are doing I calculations to select a beam, always in the table we are utilizing I_{xx} or in some books this is called I_x. In the table provided, we are simply calling it "I."**

In beam design problems, you will usually find the maximum bending stress and maximum shearing stress by first sketching the shear and moment diagram. If the material selected is metal, typically the bending stress will be the driving factor in the design. Thus, you solve for the minimum required section modulus through the equation $S = \dfrac{M_{max}}{\sigma_{allowable}}$. Once you select a beam with a S that is greater than or equal to the one required from your calculation, you will then verify that the allowable shear stress isn't exceeded.

Note: In the table, there will be more than one beam that meets the required minimum S value, so it is often necessary to find the beam that is equal to or greater than the required S and then select the one with the lightest weight/length. For steel beams, the cost is proportional to weight, therefore keeping cost down while meeting the design criteria is important.

When designing beams with wood, typically the shear stress is the driving influence on the design. Thus, start with solving for transverse shear first and then make sure your design can survive the bending stress.

DESIGN OF BEAMS AND SHAFTS

Typical wide flange I-beam geometric properties from AISC Shapes Database V16.0.

DESIGN OF BEAMS AND SHAFTS

Name	S (in³)	W (lb/ft)	d (in)	b_f (in)	t_w (in)	t_f (in)	I (in⁴)
W21X44	81.6	44.0	20.7	6.50	0.350	0.450	843
W18X65	117	65.0	18.4	7.59	0.450	0.750	1,070
W18X60	108	60.0	18.2	7.56	0.415	0.695	984
W18X55	98.3	55.0	18.1	7.53	0.390	0.630	890
W18X50	88.9	50.0	18.0	7.50	0.355	0.570	800
W18X46	78.8	46.0	18.1	6.06	0.360	0.605	712
W14X68	103	68.0	14.0	10.0	0.415	0.720	722
W14X61	92.1	61.0	13.9	10.0	0.375	0.645	640
W14X53	77.8	53.0	13.9	8.06	0.370	0.660	541
W14X48	70.2	48.0	13.8	8.03	0.340	0.595	484
W14X43	62.6	43.0	13.7	8.00	0.305	0.530	428
W14X38	54.6	38.0	14.1	6.77	0.310	0.515	385
W14X34	48.6	34.0	14.0	6.75	0.285	0.455	340
W14X30	42.0	30.0	13.8	6.73	0.270	0.385	291
W14X26	35.3	26.0	13.9	5.03	0.255	0.420	245
W14X22	29.0	22.0	13.7	5.00	0.230	0.335	199
W12X87	118	87.0	12.5	12.1	0.515	0.810	740
W12X79	107	79.0	12.4	12.1	0.470	0.735	662
W12X72	97.4	72.0	12.3	12.0	0.430	0.670	597
W12X30	38.6	30.0	12.3	6.52	0.260	0.440	238
W12X26	33.4	26.0	12.2	6.49	0.230	0.380	204
W12X22	25.4	22.0	12.3	4.03	0.260	0.425	156
W12X19	21.3	19.0	12.2	4.01	0.235	0.350	130
W12X16	17.1	16.0	12.0	3.99	0.220	0.265	103
W12X14	14.9	14.0	11.9	3.97	0.200	0.225	88.6
W10X100	112	100	11.1	10.3	0.680	1.12	623
W10X88	98.5	88.0	10.8	10.3	0.605	0.990	534
W10X19	18.8	19.0	10.2	4.02	0.250	0.395	96.3
W10X17	16.2	17.0	10.1	4.01	0.240	0.330	81.9
W10X15	13.8	15.0	9.99	4.00	0.230	0.270	68.9
W10X12	10.9	12.0	9.87	3.96	0.190	0.210	53.8
W8X35	31.2	35.0	8.12	8.02	0.310	0.495	127
W8X31	27.5	31.0	8.00	8.00	0.285	0.435	110
W8X28	24.3	28.0	8.06	6.54	0.285	0.465	98.0

EXAMPLE: WIDE FLANGE BEAM DESIGN

A steel beam has an allowable bending stress of 24 ksi and an allowable shear stress of 14.5 ksi. Find an appropriate wide flange beam for the loading scenario shown.

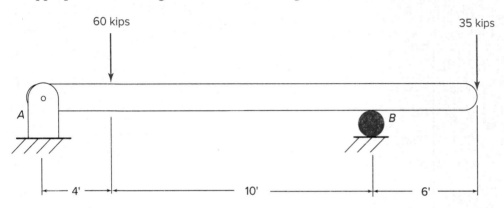

STEP 1: Find global equilibrium.

$$\Sigma M_A = 0 = -60(4) - 35(20) + B_y(14)$$

$$B_y = 67.14 \text{ kips}$$

$$\Sigma F_y = 0 = A_y + 67.14 - 60 - 35$$

$$A_y = 27.86 \text{ kips}$$

STEP 2: Compute the shear and moment diagram to find max V and max M.

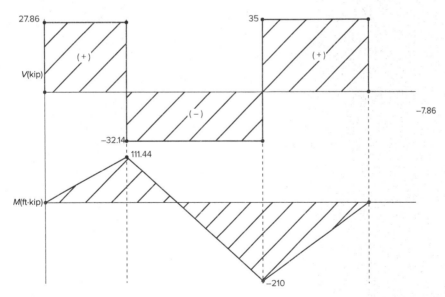

$$V_{max} = 35 \text{ kips}$$

$$M_{max} = 210 \text{ kip·ft}$$

STEP 3: Use the section modulus equation to determine S.

$$S = \frac{M_{max}}{\sigma_{allowable}} = \frac{(210 \text{ kip} \cdot \text{ft})(12 \text{ in/ft})}{24 \text{ kips/in}^2} = 105 \text{ in}^3$$

STEP 4: Look at the wide flange beam tables and use S from step 3 to select a beam.

Be careful as there will be more than one beam shape that will meet the criteria.

$W18 \times 60$ $S = 108 \text{ in}^3$

$W12 \times 87$ $S = 118 \text{ in}^3$

$W10 \times 100$ $S = 112 \text{ in}^3$

Select the $W18 \times 60$ beam as it is the lightest beam per foot (cheapest that will do the job) at 60 lb/ft.

Press pause on video lesson 61 once you get to the workout problem. Only press play if you get stuck.

A wide flange beam has a $\sigma_{allowable}$ = 22 ksi and $\tau_{allowable}$ = 12 ksi. Select the lightest wide flange beam with the shallowest depth.

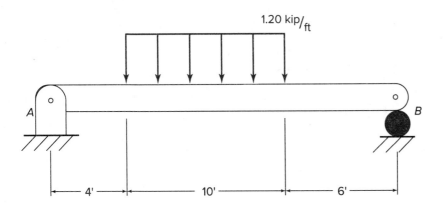

TEST YOURSELF 11.2

SOLUTION TO TEST YOURSELF: Design of Beams and Shafts

For the beam shown, determine the lightest wide flange beam while keeping the depth below 12 inches. The allowable bending stress is 22 ksi and the allowable shear stress is 14 ksi.

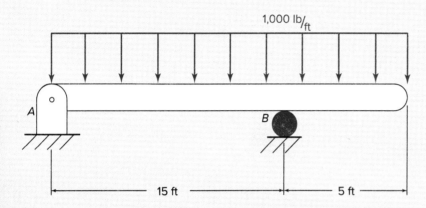

DESIGN OF BEAMS AND SHAFTS

ANSWER
W12 × 14

To select a beam that can handle the designated load use the following steps:

- Depending on the material and design codes, you will usually be working with an allowable normal and shear stress (in later courses, you may learn how to work directly with design codes to determine the allowable values).
- Create a shear and moment diagram (to find M_{max} and V_{max})
- Calculate the second moment area of inertia for the cross section of your beam
- Calculate the maximum normal and shear stress due to bending:

  - $\sigma = \dfrac{Mc}{I}$

  - $\tau = \dfrac{VQ}{It}$

  - To help make calculating the τ faster there are two possible shortcuts:

    - For a rectangular beam: $\tau = 1.5 \left(\dfrac{V}{A}\right)$

    - For an I-beam: $\tau = \left(\dfrac{V}{A}\right)$, where A is the web's depth times its thickness

- If you are using the yield stress as the design limit, you can determine the FoS of your beam by taking the material property and dividing it by the actual stress in your beam.

Often you are to determine the correct beam size. In that case, we use the bending stress equation $\sigma = \left|\dfrac{Mc}{I}\right|$ and lump all the geometry terms into one unknown called S, the section modulus, where $S = \dfrac{M_{max}}{\sigma_{allowable}}$. When you select a beam, you are paying for steel by the lb. Thus, you will want to select the lightest weight that can safely hold the specified load.

Equations for Beams and Stress

$$S = \frac{M_{max}}{\sigma_{allowable}}$$

$$S = \frac{I}{c}$$

PRO TIPS

Beam Design

- It is usually desired to find the maximum compressive stress and maximum tensile stress in a beam. Some materials behave differently in tension than in compression (i.e., concrete, which is terrible in tension but good in compression).

Shear Stress Shortcuts

- When you determine S, don't forget that it is in the denominator of the bending stress equation. Thus, you will want to pick an S on the lookup tables that is equal to or larger than the one calculated.
- Note when you purchase lumber, you are purchasing dressed lumber, which has $\frac{1}{4}$" faced off each side so it's not actually 2 x 4 but rather 1.5 inches x 3.5 inches.

Level 12

Deflection of Beams and Shafts

Slope and Deflection

Tricks and Tips

> **WATCH VIDEO** **MECHANICS OF MATERIALS LESSON 62**
> **Slope and Deflection Beam Bending Introduction**

In Statics class we learned how to solve for the deflection of a beam undergoing an axial load. Now we are going to look at deflection of a beam due to a transverse load.

It is helpful to be able to visualize what the slope and deflection of a beam would look like before solving for these entities.

- Fixed connection means it is locked and can't move. At a wall there is no deflection or slope (i.e., the beam is flat).
- Pinned connections and rollers can have a slope but can't deflect at that location.

Below are beam loading scenarios. The blue line is an exaggeration of what the beam would look like after it deflects.

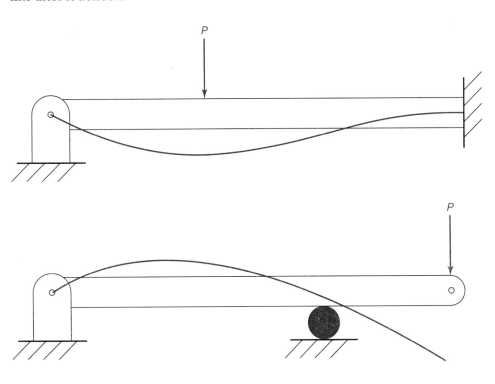

CHALLENGE QUESTION

Can you draw what the deflected beam would look like for the following?

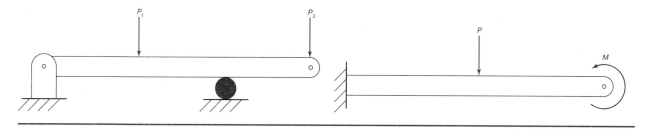

Slope is given by θ, and we refer to the displacement of the beam as "y" although other authors or books may call this "u."

the deflection.

Recall that the integral of the load curve is the shear curve and the integral of the shear curve is the moment. What we have yet to plot is the integral of the moment curve, which is the slope and the integral of the slope curve, which is

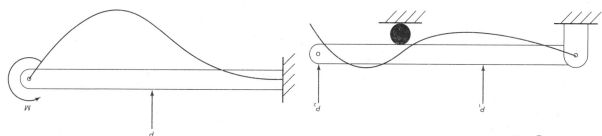

Challenge Question Solutions

✓ TEST YOURSELF 12.1

SOLUTION TO TEST YOURSELF: Deflection of Beams and Shafts

For the simply supported beam loaded as shown, sketch the shear and moment diagram. Also, sketch the deflected beam.

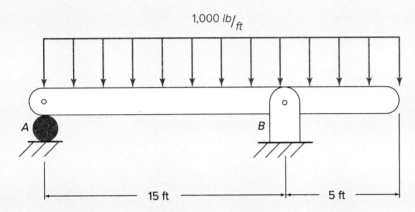

1,000 lb/ft

A

B

15 ft

5 ft

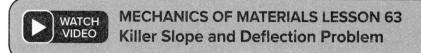

MECHANICS OF MATERIALS LESSON 63
Killer Slope and Deflection Problem

To determine the deflection of a beam, you first have to write the moment equation as a function of x (position along the length of the beam).

$$\frac{d^2y}{dx^2} = \frac{M(x)}{EI}$$

where

$M(x)$: Moment in the section of the beam as a function of x

E: Young's Modulus

I: Second moment of area or moment of inertia

When you integrate the moment equation once, you will find the slope (i.e., $\frac{dy}{dx}$). Then, when you integrate again, you will find the deflection (i.e., $y(x)$). As you should remember from calculus, when you integrate something you will get a constant of integration, "$+C$" term, that needs to be determined. This is where boundary conditions come into effect. If you can properly identify the boundary conditions, you can solve for the constants of integration.

Boundary Conditions

Boundary conditions are the absolute "knowns" on a beam (i.e., they give us the constraints that allow us to solve for constants of integration in the slope and deflection equations).

Common boundary conditions include:

■ At a wall: No slope and no deflection.

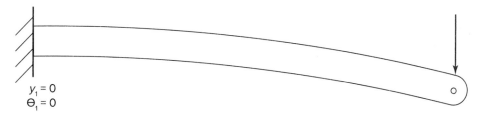

$y_1 = 0$
$\theta_1 = 0$

■   At a pin or roller: It could have a slope, so we can't so we can't assume the slope is zero, but it won't have a deflection.

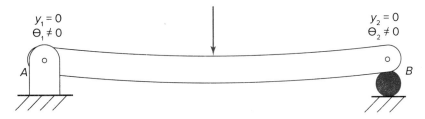

■   At the transition state between section 1 and section 2 (where the mathematical function of the load changes due to external forces being applied): Since the beam doesn't break in half, both the slopes and deflections have to be the same at that point.

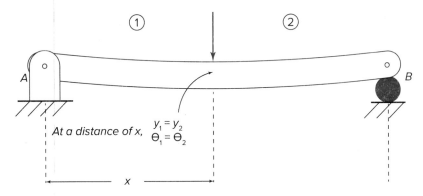

EXAMPLE: HARDER SLOPE AND DEFLECTION

Determine the slope and deflection equation for the beam between sections AB and BC.

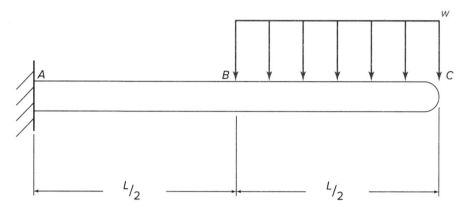

STEP 1: Find global equilibrium for the loaded beam.

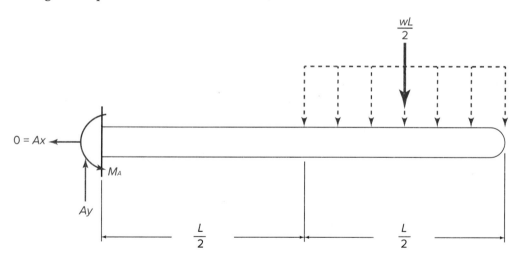

$$\Sigma F_y = 0 = A_y - (\frac{wL}{2})$$

$$A_y = (\frac{wL}{2})$$

$$\Sigma M_A = 0 = M_A - (\frac{wL}{2})(\frac{3L}{4})$$

$$\Sigma M_A = (\frac{3wL^2}{8})$$

STEP 2: Cut through the members of interest. Let's start with section AB.

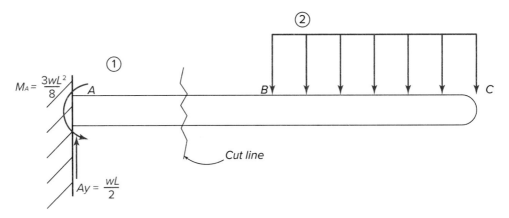

STEP 3: Draw FBD of section AB cut and find interest forces M, N, and V.

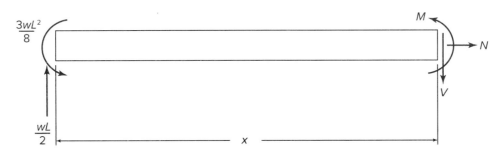

$$\Sigma M_{cut} = 0 = M + \frac{3wL^2}{8} - \frac{wL}{2}(x)$$

$$M_1 = \frac{wLx}{2} - \frac{3wL^2}{8}$$

STEP 4: Draw the FBD of section BC and find the internal forces M, N, and V.

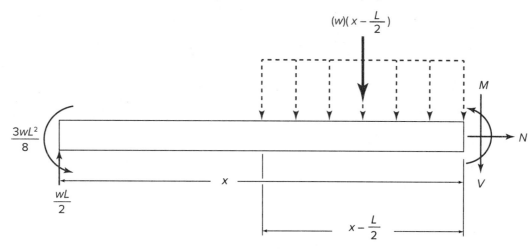

$$\Sigma M_{cut} = 0 = M_2 + \frac{3wL^2}{8} - \frac{wL}{2}(x) + (w)(x - \frac{L}{2})(x - \frac{L}{2})(\frac{1}{2})$$

$$M_2 = \frac{wLx}{2} - \frac{3wL^2}{8} - \frac{w}{2}(x^2 - Lx + \frac{L^2}{4})$$

$$M_2 = \frac{wLx}{2} - \frac{3wL^2}{8} - \frac{wx^2}{2} + \frac{wLx}{2} - \frac{wL^2}{8}$$

$$M_2 = wLx - \frac{wL^2}{2} - \frac{wx^2}{2}$$

STEP 5: Determine boundary conditions for the beam. Imagine what the deflected beam would look like. **Note:** $y =$ deflection and $\theta =$ slope of beam.

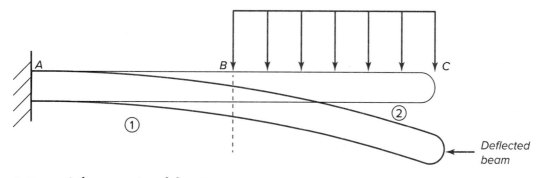

Deflected beam

$$\therefore @x_1 = 0 \text{ then } y_1 = 0 \text{ and } \theta_1 = 0$$

$$@x_1 = \frac{L}{2} \text{ then } y_1 = y_2 \text{ and } \theta_1 = \theta_2$$

STEP 6: Integrate the moment equations from step 3 using the following equation:

$$\frac{d^2y}{dx^2} = \frac{M(x)}{EI} = \frac{\dfrac{wLx}{2} - \dfrac{3wL^2}{8}}{EI}$$

First integral of the M equation yields the equation for slope.

$$\theta_1 = \frac{1}{EI}\left(\frac{wLx^2}{4} - \frac{3wL^2x}{8} + C_1\right)$$

Second integral of the M equation yields the equation for deflection.

$$y_1 = \frac{1}{EI}\left(\frac{wLx^3}{12} - \frac{3wL^2x^2}{16} + C_1x + C_2\right)$$

STEP 7: Now let's integrate the M equation for section BC of the loaded beam.

$$\frac{d^2y}{dx^2} = \frac{M(x)}{EI} = \frac{wLx - \dfrac{wL^2}{2} - \dfrac{wx^2}{2}}{EI}$$

$$\theta_2 = \frac{1}{EI}\left(\frac{wLx^2}{2} - \frac{wL^2x}{2} - \frac{wx^3}{6} + C_3\right)$$

$$y_2 = \frac{1}{EI}\left(\frac{wLx^3}{6} - \frac{wL^2x^2}{4} - \frac{wx^4}{24} + C_3x + C_4\right)$$

STEP 8: Use our boundary conditions to solve for the constant of integration.

Using $x_1 = 0$, $y_1 = 0$

$$y_1 = \frac{1}{EI}\left(\frac{wLx^3}{12} - \frac{3wL^2x^2}{16} + C_1x + C_2\right)$$

Substitute $x_1 = 0$ and $y_1 = 0$

$$0 = \frac{1}{EI}\left(\frac{wL(0^3)}{12} - \frac{3wL^2(0^2)}{16} + C_10 + C_2\right)$$

$\therefore C_2 = 0$

Using $x_1 = 0$, $\theta_1 = 0$

$$\theta_1 = \frac{1}{EI}\left(\frac{wLx^2}{4} - \frac{3wL^2x}{8} + C_1\right)$$

Substitute $x_1 = 0$ and $\theta_1 = 0$

$$0 = \frac{1}{EI}\left(\frac{wL(0^2)}{4} - \frac{3wL^2(0)}{8} + C_1\right)$$

$\therefore C_1 = 0$

DEFLECTION OF BEAMS AND SHAFTS

So,

$$\theta_1 = \frac{1}{EI}\left(\frac{wLx^2}{4} - \frac{3wL^2x}{8}\right)$$

$$y_1 = \frac{1}{EI}\left(\frac{wLx^3}{12} - \frac{3wL^2x^2}{16}\right)$$

STEP 9: Applying the boundary condition at $x = \frac{L}{2}$, $\theta_1 = \theta_2$, we can set the θ_1 equation equal to the θ_2 equation and solve for C_3.

$$\frac{wL\left(\frac{L}{2}\right)^2}{4} - \frac{3wL^2\left(\frac{L}{2}\right)}{8} = \frac{wL\left(\frac{L}{2}\right)^2}{2} - \frac{wL^2\left(\frac{L}{2}\right)}{2} - \frac{w\left(\frac{L}{2}\right)^3}{6} + C_3$$

$$C_3 = \frac{wL^3}{16} - \frac{3wL^3}{16} - \frac{wL^3}{8} + \frac{wL^3}{4} + \frac{wL^3}{48}$$

$$C_3 = \frac{wL^3}{48}$$

STEP 10: Finally, let's solve for C_4 using our final boundary condition, when $x = \frac{L}{2}$, $y_1 = y_2$.

$$\frac{wL\left(\frac{L}{2}\right)^3}{12} - \frac{3wL^2\left(\frac{L}{2}\right)^2}{16} = \frac{wL\left(\frac{L}{2}\right)^3}{6} - \frac{wL^2\left(\frac{L}{2}\right)^2}{4} - \frac{w\left(\frac{L}{2}\right)^4}{24} + \frac{wL^3\left(\frac{L}{2}\right)}{48} + C_4$$

$$C_4 = \frac{wL^4}{96} - \frac{3wL^4}{64} - \frac{wL^4}{48} + \frac{wL^4}{16} + \frac{wL^4}{384} - \frac{wL^4}{96}$$

$$C_4 = -\frac{wL^4}{384}$$

STEP 11: Write all equations in final form and simplify.

$$\theta_1 = \frac{1}{EI}\left(\frac{wLx^2}{4} - \frac{3wL^2x}{8}\right)$$

$$y_1 = \frac{1}{EI}\left(\frac{wLx^3}{12} - \frac{3wL^2x^2}{16}\right)$$

$$\theta_2 = \frac{1}{EI}\left(\frac{wLx^2}{2} - \frac{wL^2x}{2} - \frac{wx^3}{6} + \frac{wL^3}{48}\right)$$

$$y_2 = \frac{1}{EI}\left(\frac{wLx^3}{6} - \frac{wL^2x^2}{4} - \frac{wx^4}{24} + \frac{wL^3x}{48} - \frac{wL^4}{384}\right)$$

Press pause on video lesson 63 once you get to the workout problem. Only press play if you get stuck.

Find the slope and deflection at point *C* in terms of *E* and *I*.

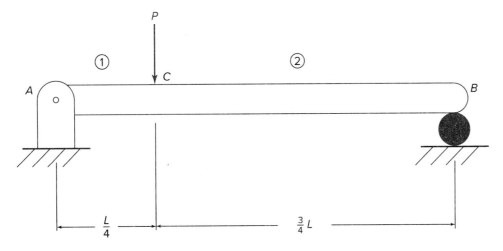

✓ TEST YOURSELF 12.2

SOLUTION TO TEST YOURSELF: Deflection of Beams and Shafts

The cantilevered beam shown has $E = 29{,}000$ ksi and $I = 140$ in^4. Compute the beam displacement at the free end of the beam. (*Hint:* Be careful of mixed units.)

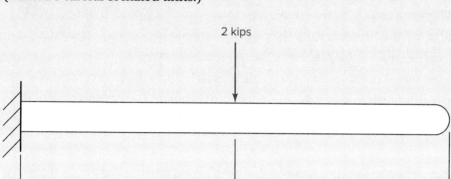

ANSWER
$y = -0.709$ in

WATCH VIDEO **MECHANICS OF MATERIALS LESSON 64**
Slope and Deflection Equation Example Problem

Whenever you can, assume x is from the left-hand side when writing the slope and deflection equations. When the beam needs to be cut multiple times to determine the entire deflection equation, you need to keep track of x. When you have x_1 and x_2, don't forget they are completely different values (in other words, x_1 is not equal to x_2) and thus will form different equations. Therefore, you need to be careful what you are plugging in (i.e., ask yourself where is the reference location?).

Distance from the right-hand side is measured negative, and distance from the left end of the beam will be positive (we have to keep our positive/negative reference frame constant). See this concept demonstrated in the following example problem.

 # EXAMPLE: SLOPE AND DEFLECTION EQUATION

Determine the slope and deflection equations of the elastic curve using x_1 and x_2 coordinates.

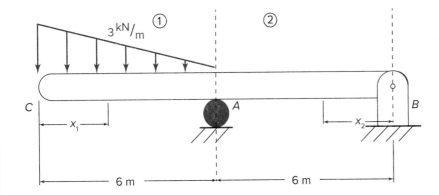

STEP 1: Find global equilibrium.

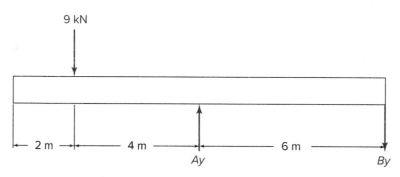

$\Sigma M_A = 0 = 9(4) - B_y(6)$

$B_y = 6 \text{ kN}$

$\Sigma F_y = 0 = A_y - 9 - 6$

$A_y = 15 \text{ kN}$

STEP 2: Cut the beam in section ①, draw the FBD and write an equation for M_1 (bending moment).

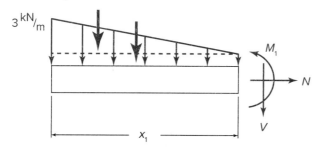

Calculate the height of the rectangle by using the slope of the line.

Slope of the line: $y = mx + b$

$$y = -\frac{3}{6}(x_1) + 3$$

$$y = 3 - \frac{x_1}{2}$$

So at $x = x_1$, $y = (3 - \frac{x_1}{2})$

which makes the area of the rectangle $(b \times h)$ equal to:

$$(x_1)(3 - \frac{x_1}{2}) = 3x_1 - \frac{x_1^2}{2}$$

and the area of the triangular portion of the load $(\frac{1}{2}bh)$ equal to:

$$\frac{1}{2}(x_1)(3 - (3 - \frac{x_1}{2})) = \frac{1}{2}(x_1)(\frac{x_1}{2}) = \frac{x_1^2}{4}$$

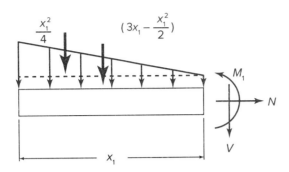

Write a moment equation about the cut to obtain the equation for M_1.

$$\Sigma M_{cut} = 0 = M_1 + (\frac{x_1^2}{4})(\frac{2}{3}x_1) + (3x_1 - \frac{x_1^2}{2})(\frac{x_1}{2})$$

$$M_1 = \frac{x_1^3}{6} - \frac{3x_1^2}{2} + \frac{x_1^3}{4}$$

$$M_1 = \frac{x_1^3}{12} - \frac{3x_1^2}{2}$$

STEP 3: Cut the beam in section ②, draw the FBD, and solve for M_2. But this time, measure from the right side of the beam.

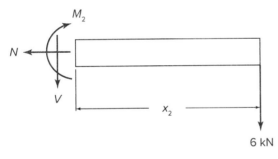

6 kN

$$\Sigma M_{cut} = 0 = -M_2 - 6(x_2)$$
$$M_2 = -6x_2$$

STEP 4: Derive your boundary conditions for the beam.

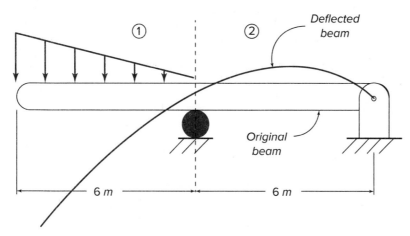

When $x_1 = 6$ m, $y_1 = 0$

When $x_2 = 0$ m, $y_2 = 0$ and when $x_2 = -6$, $y_2 = 0$

When $x_1 = 6$ m, $y_1 = y_2$ and $\theta_1 = \theta_2$

STEP 5: Integrate M_1 once to get the slope equation for section ①.

$$\theta_1 = \frac{1}{EI}\left(\int \left(\frac{x_1^3}{12} - \frac{3x_1^2}{2}\right) dx_1\right)$$

$$\theta_1 = \frac{1}{EI}\left(\frac{x_1^4}{48} - \frac{x_1^2}{2} + C_1\right)$$

Integrate once more to get the deflection equation.

$$y_1 = \frac{1}{EI}\left(\frac{x_1^5}{240} - \frac{x_1^4}{8} + C_1 x_1 + C_2\right)$$

STEP 6: Integrate M_2 once to get the slope equation for section ②.

$$\theta_2 = \frac{1}{EI}\left(\int (-6x_2) dx_2\right)$$

$$\theta_2 = \frac{1}{EI}\left(-3x_2^2 + C_3\right)$$

Integrate once more to get the deflection equation.

$$y_2 = \frac{1}{EI}\left(-x_2^3 + C_3 x_2 + C_4\right)$$

STEP 7: Use the first boundary condition to find an equation for C_2.

$y_1 = 0$ when $x_1 = 6$ (Plug into the y_1 equation $y = 0$, and $x = 6$.)

$$0 = \frac{(6)^5}{240} - \frac{(6)^4}{8} + C_1(6) + C_2$$

$$C_2 = 129.6 - 6C_1$$

STEP 8: Let's use the boundary condition for y_2. (Substitute $y_2 = 0$ and $x_2 = 0$.)

$$0 = -(0)^3 + C_3(0) + C_4$$

$$\therefore C_4 = 0$$

STEP 9: Use the boundary condition when $y_1 = y_2$, $x_1 = 6$, and $x_2 = -6$

$$\frac{x_1^5}{240} - \frac{x_1^4}{8} + C_1 x_1 + C_2 = -x_2^3 + C_3 x_2$$

Substitute in for x_1 and x_2.

$$\frac{(6)^5}{240} - \frac{(6)^4}{8} + C_1(6) + C_2 = -(-6)^3 + 6C_3$$

$$-129.6 + 6C_1 + C_2 = 216 + 6C_3$$

$$C_2 = 6C_3 - 6C_1 + 86.4$$

Substitute from step 7.

$$129.6 - 6C_1 = -6C_1 + 86.4 + 6C_3$$

$$C_3 = 7.2$$

STEP 10: Use the last boundary condition when $\theta_1 = \theta_2$, $x_1 = 6$, $x_2 = -6$

$$\frac{x_1^4}{48} - \frac{x_1^3}{2} + C_1 = -3x_2^2 + C_3$$

Substitute in for x_1 and x_2.

$$\frac{(6)^4}{48} - \frac{(6)^3}{2} + C_1 = -3(-6)^2 + 7.2$$

$$C_1 = -19.8$$

and since

$$C_2 = 129.6 - 6C_1$$

$$C_2 = 129.6 - 6(-19.8)$$

$$C_2 = 248.4$$

STEP 11: Rewrite the four equations with the constants replaced.

$$\theta_1 = \frac{1}{EI}\left(\frac{x_1^4}{48} - \frac{x_1^3}{2} - 19.8\right)$$

$$y_1 = \frac{1}{EI}\left(\frac{x_1^5}{240} - \frac{x_1^5}{8} - 19.8x_1 + 248.4\right)$$

and

$$\theta_2 = \frac{1}{EI}(3x_2^2 + 7.2)$$

$$y_2 = \frac{1}{EI}(x_2^3 + 7.2x_2)$$

PRO TIP

If you have a cantilevered beam, max deflection will always be at the free end of the beam. If you have a simply supported beam (i.e., pin connection and a roller), then max deflection will occur where the slope of the beam becomes zero! Remember calculus . . . when you take a derivative and set it equal to zero, this is the local max or min location.

Press pause on video lesson 64 once you get to the workout problem. Only press play if you get stuck.

What is the slope and deflection equation for each section of the beam?

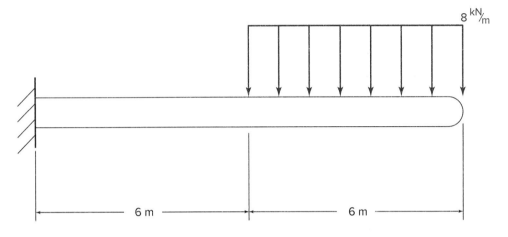

TEST YOURSELF 12.3

SOLUTION TO TEST YOURSELF: Deflection of Beams and Shafts

For the beam shown as loaded, write the equations for the displacement and slope. (**Bonus:** If $E = 29 \times 10^6$ psi and $I = 140$ in^4, what is the displacement at 10 ft?)

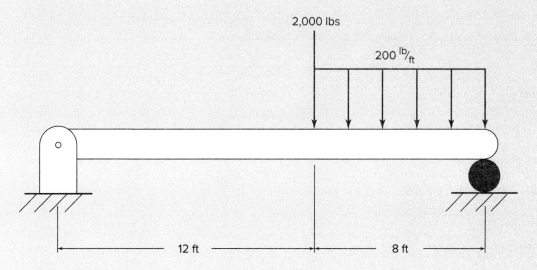

2,000 lbs

200 $^{lb}/_{ft}$

12 ft

8 ft

See solutions in the back of the book. Bonus answer is −0.195 in.

ANSWERS

DEFLECTION OF BEAMS AND SHAFTS

MECHANICS OF MATERIALS LESSON 65
Slope and Deflection Tricks and Tips to Reduce Errors!

Tips and Tricks for Solving Slope and Deflection Equations

When dimensions and loads are given in terms of letters or symbols, the slope and deflection calculations are more difficult and have higher changes of algebraic mistakes. The same problem with actual dimensions or loads can usually be solved in under half the time as those with letters or symbols. The following tips will help reduce the chances of errors.

Tip 1: When solving these problems for a value for slope and deflection, first look at the units for I (second area moment of inertia) and and then at E (modulus of elasticity). Make sure all units in the problem (i.e., dimensions and loads), match the units for I and E.

For instance, if $E = 200$ GPa and $I = 260 \times 10^6$ mm^4, then first remember 200 GPa is 200,000 MPa or $\frac{\text{N}}{\text{mm}^2}$. Now, you just need to make sure all dimensions in the problem are in mm and all loads are in N. Using this tip, you can rest assured that all units for slope will cancel out (necessary for slope in radians!) and your deflection answer will be in mm.

Tip 2: When solving for the equation for moment, you have to "cut" the beam through the section of interest. When sectioning the beam, always draw the section on the left side. This way, the distance x where the section was cut is always measured from the same point (i.e., making x in one equation the same as x in the other equation for the next section). This helps when setting one equation equal to another.

PRO TIP

For shear and moment diagrams, don't forget concentrated loads make you jump ("Van Halen" forces). Also for moments the clock is above and the counter is below for when determining which way you will jump for a coupled moment on the moment diagram. Also don't forget to "take you home." For the moment and shear diagram, you should always end back at zero for the end of the beam.

DEFLECTION OF BEAMS AND SHAFTS

EXAMPLE: EMPLOYING TIPS AND TRICKS

For the following wooden beam, $E = 12$ GPa, find the slope and the deflection at the free end.

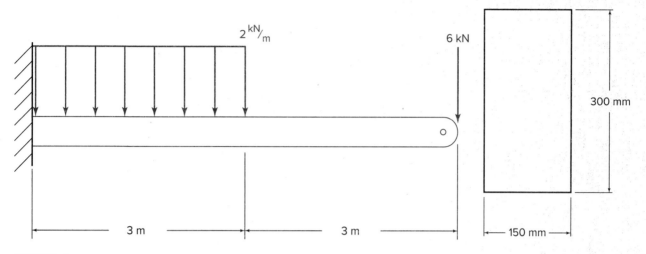

STEP 1:   Look at E and and determine the common units we are going to use.

   Note: E is in units of GPa which is kN/mm², so when calculating I, use units of mm⁴.

   Calculate I: $I = \dfrac{1}{2}(150)(300)^3 = 3.375 \times 10^8$ mm⁴

   $\therefore$ we will convert all distances to mm and use kN for all forces.

STEP 2:   Calculate global equilibrium.

$$\Sigma F_y = 0 = A_y - 6 \text{ kN} - 6 \text{ kN}$$
$$A_y = 12 \text{ kN}$$

$$\Sigma M_A = 0 = M_A - 6(1.5) - 6(6)$$
$$M_A = 45 \text{ kN·m}$$

STEP 3: Cut the beam through section 1 ($0 < x < 3$ m) and write the equation for M_1.

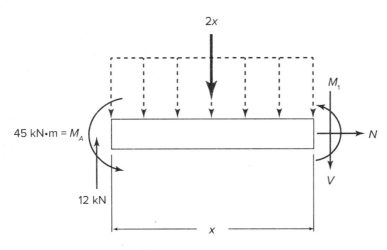

$$\Sigma M_{cut} = 0 = M_1 + 2x\left(\frac{x}{2}\right) + 45 - 12(x)$$

$$M_1 = -x^2 + 12x - 45$$

STEP 4: Cut the beam through section 2 (3 m $< x < 6$ m) and write the equation for M_2.

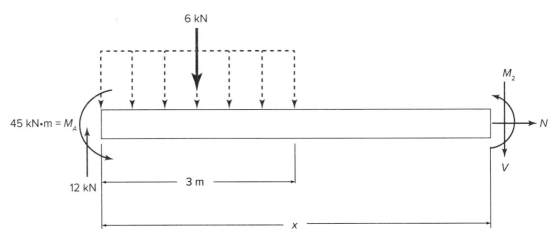

$$\Sigma M_{cut} = 0 = M_2 + 45 - 12(x) + 6(x - 1.5)$$
$$M_2 = -45 + 12x - 6x + 9$$
$$M_2 = 6x - 36$$

STEP 5: Write slope and deflection equations for section ①.

$$M_1 = -x^2 + 12x - 45$$

$$\theta_1 = \frac{1}{EI}\left(-\frac{x^3}{3} + 6x^2 - 45x + C_1\right)$$

$$y_1 = \frac{1}{EI}\left(-\frac{x^4}{12} + 2x^3 - 22.5x^2 + C_1x + C_2\right)$$

STEP 6: Write slope and deflection equations for section ②.

$$M_2 = 6x - 36$$

$$\theta_2 = \frac{1}{EI}(3x^2 - 36x + C_3)$$

$$y_2 = \frac{1}{EI}(x^3 - 18x^2 + C_3 x + C_4)$$

STEP 7: In order to solve for constants C_1, C_2, C_3, and C_4, determine boundary conditions.

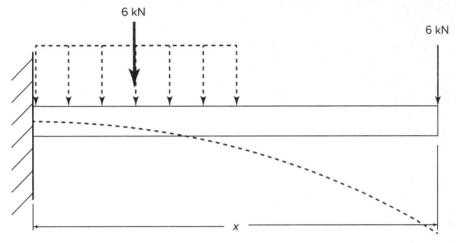

Sketch your deflected beam to visualize boundary conditions.

① When $x = 0$, $\quad \theta_1 = 0$
② And when $x = 0$, $\quad y_1 = 0$
③ Also, $x = 3{,}000$, $\quad \theta_1 = \theta_2$ $\quad\Big\}$ These equalities are correct
④ And $x = 3{,}000$, $\quad y_1 = y_2$ $\qquad$ only at the boundary
$\qquad\qquad\qquad\qquad\qquad\qquad\qquad$ between sections 1 and 2.

STEP 8: Begin solving for constants of integration using our boundary conditions. Let's start with the first boundary condition.

Take the equation for θ_1 and set $x = 0$ and $\theta_1 = 0$ and solve for C_1.

$$\theta_1 = \frac{1}{EI}\left(-\frac{x^3}{3} + 6x^2 - 45x + C_1\right)$$

$$C_1 = 0$$

STEP 9: Use the second boundary condition to solve for C_2.

Take the equation for y_1 and set $x = 0$ and $\theta_1 = 0$.

$$y_1 = \frac{1}{EI}\left(-\frac{x^4}{12} + 2x^3 - 22.5x^2 + C_1 x + C_2\right)$$

$$C_2 = 0$$

STEP 10: Use the third boundary condition to solve for C_3. Set $x = 3$.

$$-\frac{x^3}{3} + 6x^2 - 45x = 3x^2 - 36x + C_3$$

$$-\frac{(3)^3}{3} + 6(3)^2 - 45(3) = 3(3)^2 - 36(3) + C_3$$

$$-90 = -81 + C_3$$

$$C_3 = -9$$

STEP 11: Use the forth boundary condition to solve for C_4. Set $x = 3$.

$$-\frac{x^4}{12} + 2x^3 - 22.5x^2 = x^3 - 18x^2 + C_3x + C_4$$

$$-\frac{(3)^4}{12} + 2(3)^3 - 22.5(3)^2 = (3)^3 - 18(3)^2 + (-9)(3) + C_4$$

$$-155.25 = -162 + C_4$$

$$C_4 = 6.75$$

STEP 12: Find slope at the end of the beam ($x = 6$) using equation θ_2. Remember we used meters in all of our calculations deriving this equation, so we will need to match that for I and E.

$$\theta_2 = \frac{1}{EI}(3x^2 - 36x + (-9))$$

Substitute ($x = 6$).

$$\theta_2 = \frac{1}{(12\frac{kN}{mm^2})(3.375 \times 10^8\,mm^4)(\frac{m^2}{(1,000\,mm)^2})}(3(6)^2 - 36(6) - 9)$$

$$\theta_2 = -0.0289\ \text{radians}$$

STEP 13: Find deflection at the end of the beam ($x = 6$) using equation:

$$y_2 = \frac{1}{EI}(x^3 - 18x^2 + C_3x + C_4)$$

$$y_2 = \frac{1}{(12\frac{kN}{mm^2})(3.375 \times 10^8\,mm^4)(\frac{m^2}{(1,000\,mm)^2})}((6)^3 - 18(6)^2 - 9(6) + 6.75)$$

$$y_2 = -0.11833\ \text{m or } -118.33\ \text{mm}$$

PITFALL

Beware: You need to keep track of units matching! Make sure your *I* and *E* units match the units of the load and length dimension in the given problem.

▶ Press pause on video lesson 65 once you get to the workout problem. Only press play if you get stuck.

Find the max slope and deflection of the following beam.

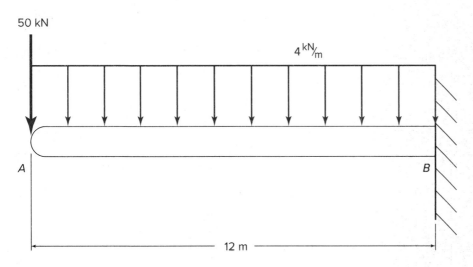

TEST YOURSELF 12.4

SOLUTION TO TEST YOURSELF: Deflection of Beams and Shafts

For the cantilevered beam with the uniform load as shown, compute the tip displacement (i.e., the free end). Take $E = 29 \times 10^6$ psi and $I = 100$ in^4.

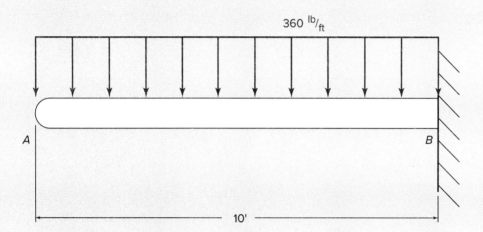

360 $^{lb}/_{ft}$

A

B

10'

ANSWER

Tip displacement = −0.268 in.

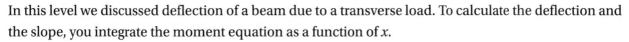

In this level we discussed deflection of a beam due to a transverse load. To calculate the deflection and the slope, you integrate the moment equation as a function of x.

$$\frac{d^2y}{dx^2} = \frac{M(x)}{EI}$$

When you integrate, you will get a constant of integration that you will need to solve for. In order to determine the value of this constant, you need to utilize the following boundary conditions:

- Fixed connection means it is locked and can't move. At a wall there is no deflection or slope (i.e., the beam is flat).
- Pinned connections and rollers can have a slope but can't deflect at that location.
- At the transition state between section 1 and section 2 (where the mathematical function of the load changes due to external forces being applied): since the beam doesn't break in half, both the slopes and deflections have to be the same at that point.

Tips and Tricks

Always find slope and deflection equations referenced from the left end of the beam when possible.

- When actually calculating a value for slope (radians) or deflection (in m or mm), be sure to have common dimensions with the values for Modulus of Elasticity (E) and the second moment of area (i.e., moment of inertia [I]).

Equations Learned in this Level

Differential Equation for Slope and Deflection

$$\frac{d^2y}{dx^2} = \frac{M(x)}{EI}$$

PRO TIPS

Slope and Deflection

- If you have a cantilevered beam, max deflection will always be at the free end of the beam. If you have a simply supported beam (i.e., pin connection and a roller) then max deflection will occur where the slope of the beam becomes zero! Remember calculus . . . when you take a derivative and set it equal to zero, this is the local max or min location.

Tricks and Tips

- For shear and moment diagrams, don't forget concentrated loads make you jump ("Van Halen" forces). Also for moments the clock is above and the counter is below for when determining which way you will jump for a coupled moment on the moment diagram. Also don't forget to "take you home." For the moment and shear diagram, you should always end back at zero for the end of the beam.

PITFALL

Tricks and Tips

- Beware: You need to keep track of units matching! Make sure your I and E units match the units of the load and length dimension in the given problem.

Level 13
Column Buckling

Column Buckling

Mission Accomplished

COLUMN BUCKLING

 WATCH VIDEO | **MECHANICS OF MATERIALS LESSON 66**
Intro to Column Buckling

Column Buckling

When designing beams, especially those members in compression, we must be mindful of buckling failures.

Buckling is a different failure type than simple yielding due to overloading in compression. Buckling is defined as a sudden change in shape of a beam or loaded member in compression resulting in beam or column failure.

There are two types of buckling failures:

1. **Short Column Buckling**
   A scaled-down example of this failure is putting weight on a soda can, resulting in "crumpling" failure, where the failure is a complete collapse.

2. **Long Column Buckling**
   The example here is applying weight to the end of a yardstick. The resulting failure is a deflection or "bowing" of the column.

So what defines a long column versus a short column?

The L/D ratio is generally what we use to determine if we have a long or short column.

L = The unsupported length of the beam or column
   (i.e., not restricting its motion)
D = The diameter or shortest thickness of the cross-section of the beam or column

If $L/D > 30$, we consider the beam or column to be a "long column." In this level we will only be looking at columns that are loaded at their centroid and are considered "long columns" such that the Euler critical buckling load is a valid equation.

Critical Load

As designers of beams or columns, it is important to know:

1. The types of supports that the column will have
2. The critical load that the column can bear before failure

Support Types. The end support conditions or "end fixity conditions" can greatly affect the load that a column can carry. Here are the end conditions you could encounter. We use K to denote end fixity conditions.

1. Pin – Pin (also called ideal conditions)

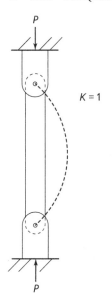

$K = 1$

2. Fixed – Fixed

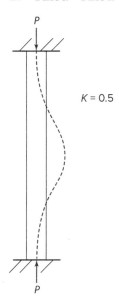

$K = 0.5$

3. Pin – Fixed

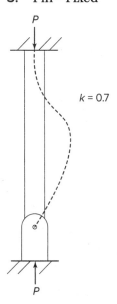

$k = 0.7$

4. Fixed – Free

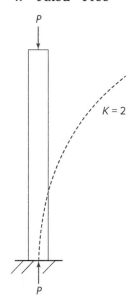

$K = 2$

COLUMN BUCKLING

Calculating the Critical Load. For finding the critical load for columns to buckle (P_{CR}), we use the Euler's Buckling Equation.

$$P_{CR} = \frac{\pi^2 EI}{(KL)^2}$$

P_{CR} = critical buckling load that the column can carry without buckling failure

E = modulus of elasticity (tells us about the flexibility of the material [material property])

I = second area moment of inertia (tells us about the "bendiness" of the beam [geometric property])

K = end fixity condition

    Pin − Pin = 1

    Fixed − Fixed = 0.5

    Fixed − Pin = 0 .7

    Fixed − Free = 2

L = unsupported (free) length of column

Note: As always, pay close attention to your units to make sure everything is the same and cancels out to leave P_{CR} in units of force (N, lbs, or kips).

Weak Versus Strong Axis of Bending

Beams or columns, typically, will have a weak and a strong axis of bending. The beam always deflects via the smaller I versus larger I.

Imagine a column constructed from an I-beam. The cross section might look like below.

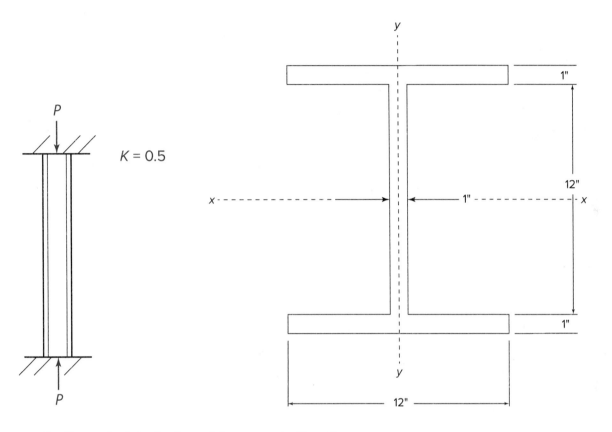

So, if we calculate the I_{xx} and I_{yy} moment of inertia,

$$I_{xx} = \frac{1}{12}(12)(14)^3 - \frac{1}{12}(11)(12)^3 = 1{,}160 \text{ in}^4$$

$$I_{yy} = \frac{1}{12}(14)(12)^3 - 2\left[\frac{1}{12}(12)(5.5)^3 + (5.5)(12)(3.25)^2\right] = 2{,}016 - 1{,}727 = 289 \text{ in}^3$$

Notice the values for the second area moment of inertia, I, are different in the I_{xx} than in the I_{yy} direction. Since I_{yy} is the smaller value, this is the weak axis of bending and is also the axis the beam will "bend around" during buckling. An I-beam will always fail around the weak axis of bending. This is the reason we call this an I-beam and not an H-beam, as it is always strongest when used in the "I" direction.

Bending Around an Axis

Often we hear "bending around the *x* axis" or "bending around the *y* axis." Imagine the "bent" or "deflected" beam forming a complete circle; if you pointed at the center of that circle, you would be pointing along the axis of bending.

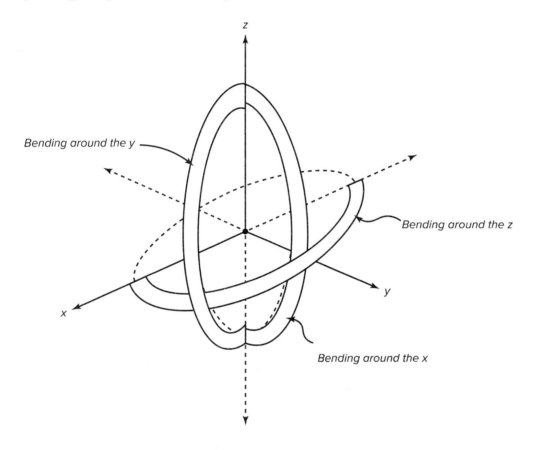

COLUMN BUCKLING

EXAMPLE: CALCULATING CRITICAL BUCKLING FORCE

The column shown is pin connected at its midsection to prevent buckling in the weak axis of bending for the column. Assume the column is ideally supported (think of it as ball and socket joints). Also note the typical (typ) comment means that the thickness of the beam is 25 mm everywhere. $E = 200$ GPa. Solve for the critical load.

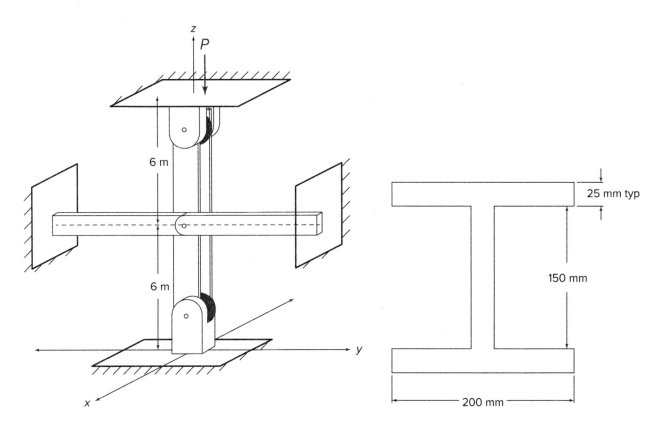

STEP 1: Calculate the critical buckling force for buckling around the x axis.

$$P_{CR} = \frac{\pi^2 EI}{(KL)^2}$$

$E = 200{,}000 \text{ MPa } \left(\frac{N}{mm^2}\right)$

$K = 1$

$L = 6{,}000 \text{ mm}$

$$I_{xx} = \frac{1}{12}(200)(200)^3 - 2\left[\frac{1}{12}(150)(87.5)^3 + (13{,}125)(56.25)^2\right]$$

$I_{xx} = 133{,}333{,}333 - 99{,}804{,}687.5$

$I_{xx} = 33{,}528{,}645.8 \text{ mm}^4$

$$P_{CR} = \frac{(\pi^2(200{,}000 \ \frac{N}{mm^2})(33{,}528{,}645.8 \text{ mm}^4)}{[(1)(6{,}000 \text{ mm})]^2} = 1{,}838.4 \text{ kN}$$

STEP 2: Calculate the critical buckling force for buckling around the y axis.

$$P_{CR} = \frac{\pi^2 EI}{(KL)^2}$$

$E = 200{,}000 \text{ MPa } \left(\frac{N}{mm^2}\right)$

$K = 1$

$L = 12{,}000 \text{ mm}$

$$I_{yy} = \frac{1}{12}(200)(200)^3 - \frac{1}{12}(175)(150)^3$$

$I_{yy} = 133{,}333{,}333 - 49{,}218{,}750$

$I_{yy} = 84{,}114{,}583.3 \text{ mm}^4$

$$P_{CR} = \frac{(\pi^2(200{,}000 \ \frac{N}{mm^2})(84{,}114{,}583.3 \text{ mm}^4)}{[(1)(12{,}000 \text{ mm})]^2} = 1{,}153 \text{ kN}$$

$\therefore$ The column will buckle around the y axis since P_{CR} is smaller, 1,153 kN < 1,838.4 kN as it will fail around the smallest P_{CR} force first.

Press pause on video lesson 66 once you get to the workout problem. Only press play if you get stuck.

Find the critical buckling load for the given 10 ft long column. End conditions are ideal. Use $E = 29 \times 10^3$ ksi.

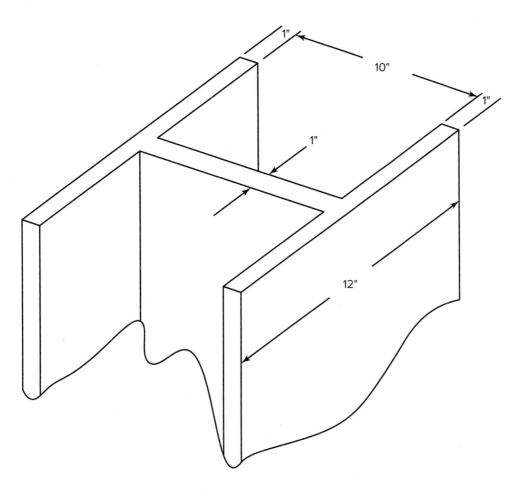

COLUMN BUCKLING

 TEST YOURSELF 13.1

SOLUTION TO TEST YOURSELF: Column Buckling

For an 8 ft wood column, with rectangular cross section of 4 in by 6 in, compare the critical buckling load for pinned-pinned and fixed-fixed boundary conditions. Take $E = 1.5 \times 10^6$ psi. (**Hint:** It will buckle in the weakest direction, i.e., the direction with the lowest I.)

WATCH VIDEO **MECHANICS OF MATERIALS LESSON 67**
Beam Column Buckling Example

Sometimes, a beam is loaded in the usual way and there is a compressive (axial) force. Such problems are called beam-columns, and their analysis is tricky! For those problems, advanced course work in solid mechanics is needed.

Although rare, beams or columns under compressive loads could also fail due to compressive yielding. To be absolutely complete in your analysis, you should always check for yielding ($\sigma = \frac{P}{A}$).

PITFALL

The common mistakes made when calculating the critical buckling are:

1. Using the the incorrect I for the axis it is buckling around (i.e., when bending around the x axis, make sure you are using I_{xx} and not I_{yy}).
2. Not using the correct end fixity conditions for the different column buckling directions (i.e., when looking at a pin, in one of the planes it is free to rotate; however, when looking at buckling in a perpendicular plane, the pin will behave as a fixed connection).

PRO TIP

Most often column buckling problems start with a statics problem. It is only the compressive component of the force that causes the beam to buckle. As an example, if you are to find buckling of members of a truss, solve only for members that are in compression as tension members won't buckle.

EXAMPLE: FACTOR OF SAFETY

Find the factor of safety of the column *BC* for buckling. Remember, a pin connection only serves as a pin in one direction. In other words, the pin at point *C* acts as a pin in one direction and a fixed support in the other direction. $E = 200$ GPa

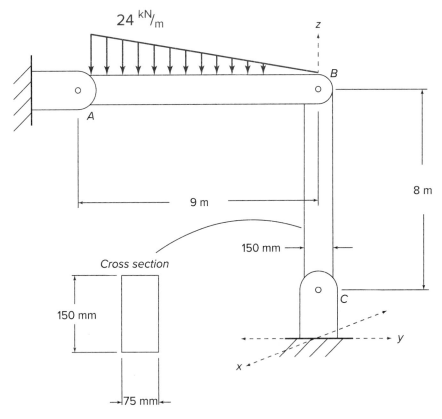

STEP 1:   Let's work a statics problem and find the forces at point *A* and point *B*.
              Draw the FBD of Bar *AB*.

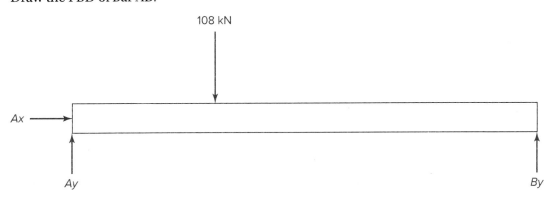

$$\Sigma M_A = 0 = B_y(9) - 108(3)$$

$$B_y = 36 \text{ kN}$$

$$\Sigma F_y = 0 = A_y + 36 - 108$$

$$A_y = 72 \text{ kN}$$

$$A_x = 0$$

So B_y is the compressive force on column BC.

STEP 2: Calculate the actual critical force that will cause the column to buckle, and remember we need to calculate both bending conditions (bending around the x and y axis).

For buckling around the x axis:

$$E = 200{,}000 \text{ MPa } (\frac{\text{N}}{\text{mm}^2})$$

$$K = 1$$

$$L = 8{,}000 \text{ mm}$$

$$I_{xx} = \frac{1}{12}(75)(150)^3 = 21{,}093{,}750 \text{ mm}^4$$

$$P_{CR} = \frac{(\pi^2(200{,}000\frac{\text{N}}{\text{mm}^2})(5{,}273{,}437.5 \text{ mm}^4)}{[(0.5)(8{,}000 \text{ mm})]^2} = 650.58 \text{ kN}$$

For buckling around the y axis:

$$E = 200{,}000 \text{ MPa } (\frac{\text{N}}{\text{mm}^2})$$

$$K = 0.5$$

$$L = 8{,}000 \text{ mm}$$

$$I_{yy} = \frac{1}{12}(150)(75)^3 = 5{,}273{,}437.5 \text{ mm}^4$$

$$P_{CR} = \frac{(\pi^2(200{,}000\frac{\text{N}}{\text{mm}^2})(5{,}273{,}437.5 \text{ mm}^4)}{[(0.5)(8{,}000 \text{ mm})]^2} = 650.58 \text{ kN}$$

$\therefore$ The critical buckling force is 650.6 kN. (Fails at the exact same force in both directions.)

STEP 3: Calculate the factor of safety.

$$\text{FoS} = \frac{\text{Force Allow}}{\text{Force Actual}} = \frac{650.58 \text{ kN}}{36 \text{ kN}} = 18.1$$

▶ Press pause on video lesson 67 once you get to the workout problem. Only press play if you get stuck.

What is the maximum distributed load that can be applied to the system with a FOS of 1.5 if the yield stress is 250 MPa and E_{st} = 200 GPa.

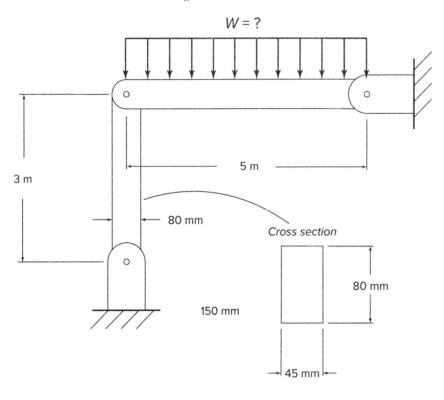

COLUMN BUCKLING

TEST YOURSELF 13.2

SOLUTION
TO TEST
YOURSELF:
Column
Buckling

Asolid circular rod, 1.5 m long, is pinned to a continuous, frictionless cable and supports an unknown load *W* as shown. If the yield stress of the rod 325 MPa with an elastic modulus of 200 GPa, at what weight will the rod fail simultaneously by buckling and yielding in compression?

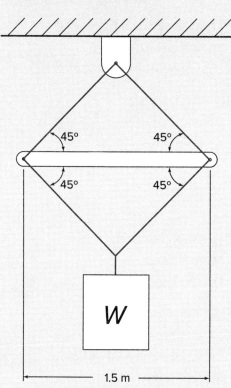

COLUMN BUCKLING

 **MECHANICS OF MATERIALS LESSON 68**
Solids Complete! What's Next?

Below is a summary of everything we covered in this mechanics of materials unbook. As you go over the list, see if you recall how to do each of these things:

- Calculating Stress
  - Shear Stress (τ)
  - Normal Stress (σ)

- Calculating Strain
  - Shear Strain (γ)
  - Normal Strain (ε)

- Determining the Mechanical Properties of Materials
  - Elastic Modulus (Young's Modulus) (E)
  - Shear Modulus (G)
  - Poisson's Ratio (ν)
  - 0.2% Offset Rule
  - Yield Stress (Y)

- Axial Elongation
  - Elongation Due to Load (δ)
  - Elongation Due to Temperature (δ_T)
  - Stress Concentrations

- Torsion
  - Shear from Torsion (τ)
  - Angle of Twist (θ)
  - Power Transmission (P)

- Beam Bending
  - Flexure Formula (σ)
  - Unsymmetric Bending
  - Composite Beams

- Transverse Shear
  - Transverse Shear Stress (τ)
  - Shear Flow (q)
  - First Area Moment of Inertia (Q)

- Combined Loadings
  - Pressure Vessels
    - Hoop Stress (σ_H)
    - Longitudinal Stress (σ_L)
    - Effect Maps

- Stress Transformation
  - Equation Method
  - Mohr's Circle
  - Failure Theories

- Strain Transformation
  - Strain Transformations with Equations and Mohr's Circle
  - Strain Gauge
  - Strain Rosettes

- Beam Bending—Slope and Deflection
  - Slope and Deflection Equations (y and θ)

- Design of a Beam
  - Section Modulus (S)

- Column Buckling
  - Euler's Buckling Load (P_{CR})
  - Column End Conditions

- Buckling is a different failure type than simple yielding due to overloading in compression.
- If the unsupported length of the beam or column (i.e., not restricting its motion) divided by the diameter or shortest thickness of the cross section of the beam or column is greater than 30, the beam or column is considered "long."
- If the long column is loaded at its centroid, the Euler critical buckling load is a valid equation.

  - $$P_{CR} = \frac{\pi^2 EI}{(KL)^2}$$

  - P_{CR} = critical buckling load that the column can carry without buckling failure (don't forget to check your units).

  - I is the second area moment of inertia. Make sure you have the correct I paired with the correct end fixity condition.

  - K depends on the end fixity conditions:
    - $K = 2$ for fixed-free
    - $K = 1$ for pinned-pinned
    - $K = 0.7$ for fixed-pinned
    - $K = 0.5$ for fixed-fixed

PRO TIP

Column Buckling

Most often column buckling problems start with a statics problem. It is only the compressive component of the force that causes the beam to buckle. As an example, if you are to find buckling of members of a truss, solve only for members that are in compression as tension members won't buckle.

PITFALL

Column Buckling

The common mistakes made when calculating the critical buckling are:

1. Using the the incorrect I for the axis it is buckling around (i.e., when bending around the x axis, make sure you are using I_{xx} and not I_{yy}) .
2. Not using the correct end fixity conditions for the different column buckling directions (i.e., when looking at a pin, in one of the planes it is free to rotate; however, when looking at buckling in a perpendicular plane, the pin will behave as a fixed connection).

Test Yourself
Solutions

Test Yourself 1.1

DISTANCE

$$2\,ft \Rightarrow mm \qquad 2\,ft \cdot \frac{12\,in}{1\,ft} \cdot \frac{25.4\,mm}{1\,in} = \boxed{609.6\,mm}$$

FORCE

$$170\,lbs \Rightarrow kg \qquad 170\,lb \cdot \frac{0.454\,kg}{1\,lb} = \boxed{77.2\,kg}$$

UNIT WEIGHT

$$150\,lb/FT^3 \Rightarrow kg/m^3 \qquad 150\,\frac{lb}{FT^3} \cdot \frac{0.454\,kg}{1\,lb} \cdot \frac{(1\,FT)^3}{(12\,in)^3} \cdot \frac{(1\,in)^3}{(0.0254\,m)^3}$$

$$= \boxed{2,405\,kg/m^3}$$

PRESSURE

$$32\,psi \Rightarrow Pa \qquad 32\,\frac{lb}{in^2} \cdot \frac{6895\,Pa}{1\,lb/in^2} = 220,640\,Pa$$

OR $\boxed{220.6\,kPa}$

ROTATION

$$2,500\,RPM \Rightarrow \underline{radians} \qquad 2,500\,\frac{Rev}{min} \cdot \frac{1\,min}{60\,s} \cdot \frac{2\pi\,radians}{1\,Rev} = \boxed{261.8\,\frac{radians}{s}}$$

Test Yourself 1.2

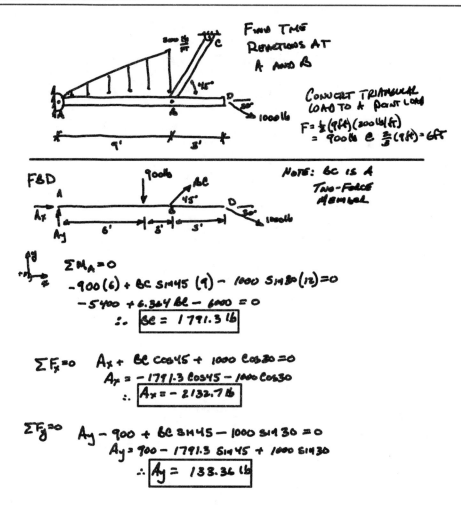

Find the reactions at A and B.

Convert triangular load to a point load

$F = \frac{1}{2}(9ft)(200\,lb/ft)$
$= 900\,lb \ @ \ \frac{2}{3}(9ft) = 6ft$

FBD

Note: BC is a Two-Force Member

$\sum M_A = 0$
$-900(6) + BC\,\sin45\,(9) - 1000\,\sin30\,(12) = 0$
$-5400 + 6.364\,BC - 6000 = 0$
$\therefore \boxed{BC = 1791.3\,lb}$

$\sum F_x = 0 \quad A_x + BC\cos45 + 1000\cos30 = 0$
$A_x = -1791.3\cos45 - 1000\cos30$
$\therefore \boxed{A_x = -2132.7\,lb}$

$\sum F_y = 0 \quad A_y - 900 + BC\sin45 - 1000\sin30 = 0$
$A_y = 900 - 1791.3\sin45 + 1000\sin30$
$\therefore \boxed{A_y = 133.36\,lb}$

Find M, V, N at a location 4' to the right of Pin A.

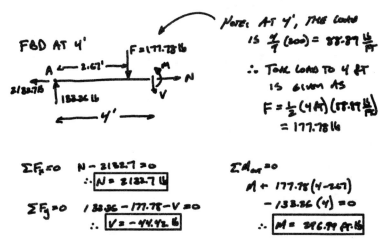

FBD at 4'

Note: At 4', the load is $\frac{4}{9}(200) = 88.89\,\frac{lb}{ft}$

$\therefore$ Total load to 4 ft is given as
$F = \frac{1}{2}(4ft)(88.89\,\frac{lb}{ft})$
$= 177.78\,lb$

$\sum F_x = 0 \quad N - 2132.7 = 0$
$\therefore \boxed{N = 2132.7\,lb}$

$\sum F_y = 0 \quad 133.36 - 177.78 - V = 0$
$\therefore \boxed{V = -44.42\,lb}$

$\sum M_{cut} = 0$
$M + 177.78(4 - 2.67)$
$- 133.36(4) = 0$
$\therefore \boxed{M = 296.99\,ft\cdot lb}$

Test Yourself 1.3

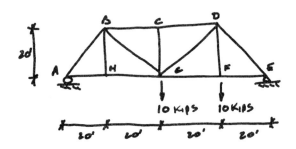

COMPUTE THE NORMAL STRESS
IN MEMBERS GF, GD, CD

Member	Area
GF	$5 in^2$
GD	$3 in^2$
CD	$6 in^2$

1) Compute the Reaction at E

$\Sigma F_x = 0 \therefore \boxed{E_x = 0}$

$\Sigma M_A = 0 \quad -10(40) - 10(60) + E_y(80) = 0$

$\therefore \boxed{E_y = 12.5 \, kips}$

Since we are only looking for the forces in three members, we will use the Method of Sections.

2) Cut through members as needed. All members are assumed to be in tension (note forces pointing away from the cut). Therefore, a negative result means the member is in compression.

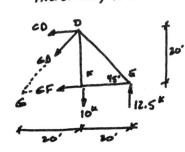

$\Sigma M_G = 0 \quad CD(20) - 10(20) + 12.5(40) = 0$

$\therefore \boxed{CD = -15.0 \, kips} \, C$

$\Sigma F_y = 0 \quad 12.5 - 10 - GD \cos 45 = 0$

$\therefore \boxed{GD = 3.54 \, kips} \, T$

$\Sigma M_D = 0 \quad 12.5(20) - GF(20) = 0$

$\therefore \boxed{GF = 12.5 \, kips} \, T$

3) Compute the Normal Stress: $\sigma_n = Force/Area$

Member	Area	Force	σ_n
CD	$6 in^2$	15 kips	2,500 psi C
GD	$3 in^2$	3.54 kips	1,180 psi T
GF	$5 in^2$	12.5 kips	2,500 psi T

Test Yourself 1.4

(1) find k

$$\Sigma F_x = 0 \quad -50,000 + \int_0^{25} k x^2 \, dx = 0$$

$$50,000 = \tfrac{1}{3} k x^3 \Big|_0^{25}$$

$$50,000 = \tfrac{1}{3} k (25)^3$$

$$\therefore \boxed{k = 9.6 \ \text{lb/ft}^3}$$

(2) FBD AT LOCATION "x"

$$\Sigma F_x = 0$$

$$-F + \int_x^{25} 9.6 \, \eta^2 \, d\eta = 0 \qquad \text{DUMMY VARIABLE USED SINCE } x \text{ IS USED AS THE LIMIT}$$

$$F = \int_x^{25} 9.6 \, \eta^2 \, d\eta = \frac{9.6}{3} \eta^3 \Big|_x^L$$

$$F = 3.2 \, (25^3 - x^3)$$

$$F = 50,000 - 3.2 \, x^3 \ \text{lb}$$

$$\therefore$$

$$\boxed{\sigma_n = \frac{F}{A} = \frac{F}{10 \text{in}^2} = 5000 - 0.32 x^3 \ \text{lb/in}^2}$$

NOTE: WE USED x IN ft

Test Yourself 1.6

(A)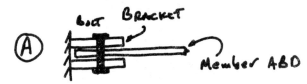

TOP VIEW

SHOWING THE BOLT LOCATION
"DOUBLE SHEAR"

SIDE VIEW

BOLT DIA = $3/8"$

$\therefore$ AREA $= \dfrac{\pi (0.375)^2}{4}$

$= 0.1104$ in^2

REACTIONS AT A FROM Problem 1.2

2132.7 lb

133.36

MAGNITUDE $= \sqrt{(2132.7)^2 + (133.36)^2}$

$\therefore$ | SHEAR FORCE $= 2136.9$ lb |

AVG SHEAR STRESS, DOUBLE SHEAR $= \dfrac{\text{SHEAR FORCE}}{2A}$

$\tau = \dfrac{2136.9 \text{ lb}}{2(0.1104 \text{ in}^2)} = 9,678$ PSI

(B) P_{IN} B IS IN SINGLE SHEAR ; BOLT DIA $= 3/8"$

SHEAR FORCE $= 1791.3$ lb (FROM Problem 1.2)

$\tau = \dfrac{F}{A} = \dfrac{1791.3 \text{ lb}}{0.1104} = 16,226$ psi

Test Yourself 1.7

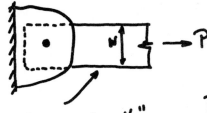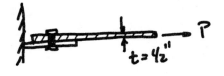

PLATE $t = 1/2"$
$w = 10"$

$P = 7.5\,KIPS$ BOLT DIAMETER $= 1.00"$

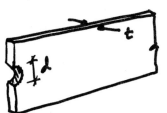

BEARING AREA $= d \times t$
PER BOLT
$= (1.00")(1/2") = 0.500\,in^2$

LOAD PER BOLT $= 7.5\,kips$

$\therefore$ BEARING STRESS $= \dfrac{7,500\,lb}{0.500\,in^2}$
$= 15,000\;psi$

NET AREA of PLATE $=$
$A_{net} = (w - d)t$
$= (10" - 1")\,1/2"$
$A_{net} = 4.5\,in^2$

$\therefore$ $\sigma_n = \dfrac{7,500\,lb}{4.5\,in^2}$
$= 1,667\;psi$

Test Yourself 1.8

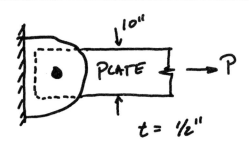

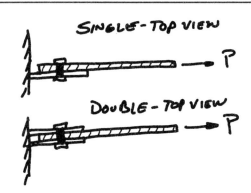

Ⓐ   ∴ $F_{BOLT} = 7.5^{KIPS}$

ALSO, WE HAVE $F.S. = \dfrac{\tau_{Allowable}}{\tau_{ACTUAL}} \Rightarrow 1.75 = \dfrac{15{,}000\ psi}{\tau_{ACT.}}$

Ⓑ SINGLE SHEAR

$\tau_{ACT} = \dfrac{7500\ lb}{A_{BOLT}}$ ∴ $1.75 = \dfrac{15{,}000\ psi}{7500 lb/A_{BOLT}}$

SOLVE FOR A_{BOLT} ∴ $A_{BOLT} = \dfrac{1.75(7500)}{15000} = 0.875\ in^2 = \dfrac{\pi D^2}{4}$

∴ $D = 1.06\ in$

∴ USE $D = 1.25\ in$ BOLT

Ⓒ DOUBLE SHEAR

$\tau_{ACT} = \dfrac{7500 lb}{2 A_{BOLT}} \Rightarrow A_{BOLT} = \dfrac{1.75(7500)}{2(15{,}000)} = 0.4375\ in^2$

$0.4375\ in^2 = \pi D^2/4$ ∴ $D = 0.7464\ in$

∴ USE $D = 3/4"$ BOLT

Test Yourself 2.1

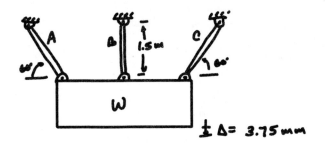

$\pm \Delta = 3.75\,mm$

Rod B: $\quad L_b = 1.5\,mm$
$\qquad\quad \Delta L = 0.00375\,m$

$\therefore \boxed{\epsilon_B = \dfrac{0.00375}{1.5} = 0.00250}$

Rod A ; C

Closer Look:

$0.00375\,M$

$\cos 30 = \dfrac{1.5}{L_A}$

$L_A = 1.732\,m$

L_A

$\Delta_A \{ \; 0.00375$

$30°$

Δ_A

$\therefore \cos 30° = \dfrac{\Delta_A}{0.00375}$

$\therefore \Delta_A = 0.00325\,M$

$\therefore$ Rod A ; C $\begin{array}{l} L_A = 1.732\,m \\ \Delta_A = 0.00325 \end{array} \Bigg\}$ $\boxed{\epsilon_A = \epsilon_c = \dfrac{0.00325}{1.732} = 0.00188}$

Test Yourself 2.2

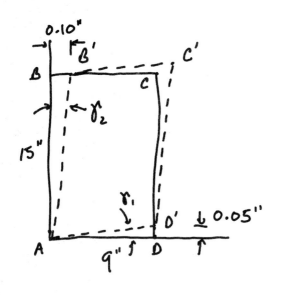

Uniform Deformation
Strain at A?

Find γ_1 ; γ_2

$$\tan\gamma_1 = \frac{0.05"}{9"}$$

$$\therefore \gamma_1 = 0.00556 \text{ Radian}$$

$$\tan\gamma_2 = \frac{0.10"}{15"}$$

$$\gamma_2 = 0.00667 \text{ Radian}$$

Strain at A

$$\gamma_A = \gamma_1 + \gamma_2 = 0.00556 + 0.00667$$

$$= \boxed{0.0122 \text{ Radians}}$$

Test Yourself 3.2

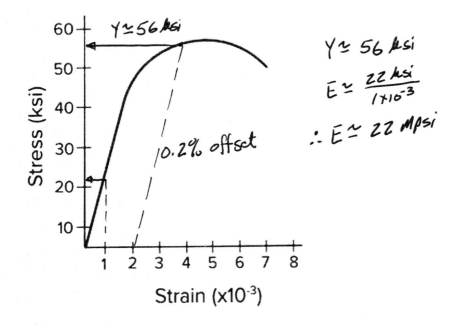

$Y \approx 56$ ksi

$E \approx \dfrac{22 \text{ ksi}}{1\times10^{-3}}$

$\therefore E \approx 22$ MPSI

Test Yourself 3.3

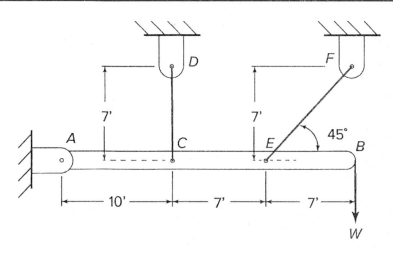

$$\frac{\Delta_C}{10\,ft} = \frac{0.07\,in}{24\,ft} \qquad \Delta_C = (0.07\,in)\,\frac{10\,ft}{24\,ft} = 0.02917\,in$$

$$\therefore \Delta_C = 0.02917\,in$$

$$\boxed{\varepsilon_{CD} = \frac{\Delta_C}{L_{CD}} = \frac{0.02917\,in}{(7 \times 12)\,in} = 3.473 \times 10^{-4}\,\frac{in}{in}}$$

$$\sigma_{CD} = E\,\varepsilon_{CD} = \left(29 \times 10^{6}\,\frac{lb}{in^{2}}\right)\left(3.473 \times 10^{-4}\,in/in\right)$$

$$\sigma_{CD} = 10{,}070\,lb/in^{2}$$

$$F_{CD} = \sigma_{CD} \cdot A = 10{,}070\,\frac{lb}{in^{2}}\,(1.5\,in^{2})$$

$$\boxed{F_{CD} = 15{,}110 \ lb}$$

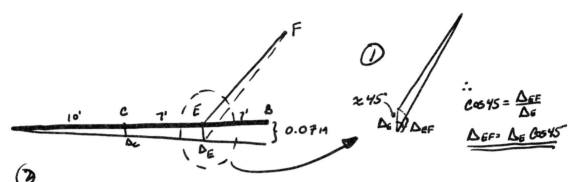

① $\therefore \cos 45 = \dfrac{\Delta_{EF}}{\Delta_E}$

$\underline{\Delta_{EF} = \Delta_E \cos 45}$

② $\therefore \Delta_{EF} = \Delta_E \cos 45$

$$\dfrac{\Delta_E}{17'} = \dfrac{0.07\text{ in}}{24'}$$

$\Delta_E = 0.07\text{ in}\left(\dfrac{17}{24}\right) = 0.04958\text{ in}$

$\therefore \Delta_{EF} = \Delta_E \cos 45$

$\qquad = 0.04958\text{ in} \cos 45$

$\qquad \Delta_{EF} = 0.03506\text{ in}$

③ $\varepsilon_{EF} = \dfrac{\Delta_{EF}}{L_{EF}} = \dfrac{0.03506\text{ in}}{(7 \times 12)\text{in} / \cos 45}$

$\boxed{\varepsilon_{EF} = 2.951 \times 10^{-4}\text{ in/in}}$

$\sigma_{EF} = E\varepsilon_{EF} = 29 \times 10^6\text{ psi }(2.951 \times 10^{-4}\text{ in/in})$

$\sigma_{EF} = 8{,}558\text{ psi}$

$F_{EF} = \sigma_{EF} \cdot A_{EF} = 8{,}558\text{ psi }(1.5\text{ in}^2)$

$\boxed{F_{EF} = 12{,}840\text{ lb}}$

FBD

④ $\sum M_A = 0$

$(15{,}110\text{ lb})10\text{ft} + 12{,}840(17\text{ft})\sin 45$

$\qquad - W(24\text{ft}) = 0$

$\therefore \boxed{W = 12{,}730\text{ lb}}$

Test Yourself 3.4

$\Delta = 0.029171\,in$ (FROM TEST YOURSELF 3.3)

$L = 7ft \times 12 = 84\,in$

$\varepsilon_{long} = \dfrac{\Delta L}{L} = \dfrac{0.029171\,in}{84\,in} = 347.3 \times 10^{-6}\,in/in$

SINCE $\nu = -\dfrac{\varepsilon_{DIA}}{\varepsilon_{long}} = 0.32$

$A = 15\,in^2$

$A = \dfrac{\pi D^2}{4}$

$\varepsilon_{DIA} = (-0.32)(347.3 \times 10^{-6})$

$= -111.1 \times 10^{-6}\,in/in$

$D = \sqrt{\dfrac{4A}{\pi}} = 1.382\,in$

$\varepsilon_{DIA} = \dfrac{\Delta D}{D_o} \Rightarrow \Delta D = \varepsilon_{DIA}(1.382\,in)$

$\therefore \Delta D = -111.1 \times 10^{-6}\,in/in\,(1.382\,in)$

$\boxed{\Delta D = -0.000154\,in}$

THIS CHANGE IS TYPICALLY NEGLECTED.

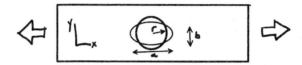

NOTE THAT THE ORIGINAL "LENGTH" OF THE CIRCLE IN x AND y IS:

$$L_x = 2r \qquad L_y = 2r$$

THE CHANGE IN LENGTH IS THEREFORE:

$$\Delta_x = a - 2r \qquad \Delta_y = b - 2r$$

SO WE CAN WRITE THE STRAIN AS:

$$\varepsilon_{xx} = \dfrac{\Delta_x}{L_x} = \dfrac{a - 2r}{2r}$$

$$\varepsilon_{yy} = \dfrac{\Delta_y}{L_y} = \dfrac{b - 2r}{2r}$$

POISSON'S RATIO FOR THIS CASE IS:

$$\nu = -\dfrac{\varepsilon_{yy}}{\varepsilon_{xx}} = -\dfrac{(b - 2r)/2r}{(a - 2r)/2r}$$

$$\therefore \boxed{\nu = -\left(\dfrac{b - 2r}{a - 2r}\right)}$$

If $r = 1.00\,in$ AND THE CIRCLE ELONGATES BY $0.2\,in$ IN THE x-DIRECTION AND DECREASES BY $0.06\,in$ IN THE y-DIRECTION, WHAT IS ν?

LET $\left.\begin{array}{l} a = 2.2'' \\ b = 1.94'' \\ r = 1.00'' \end{array}\right\}$ $\nu = -\left(\dfrac{1.94 - 2(1)}{2.2 - 2(1)}\right)$

$$\therefore \boxed{\nu = 0.30}$$

Test Yourself 3.5

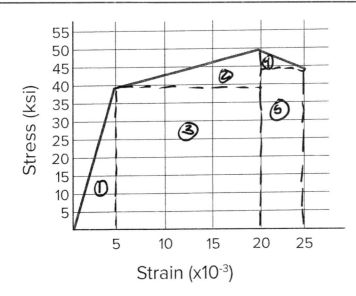

ELASTIC Modulus: $E \simeq \dfrac{40,000 \, PSi}{5 \times 10^{-3}} \simeq 8 \, MPSi$

Modulus of Resilience: Area ①

$\simeq \frac{1}{2}\left(5 \times 10^{-3}\right)\left(40,000 \, PSi\right) = 100 \, PSi$

Modulus of Toughness: All Areas

$\simeq \frac{1}{2}\left(5 \times 10^{-3}\right)\left(40,000 \, PSi\right) + \frac{1}{2}\left(15 \times 10^{-3}\right)\left(10,000 \, PSi\right)$
$+ \left(15 \times 10^{-3}\right)\left(40,000 \, PSi\right) + \frac{1}{2}\left(5 \times 10^{-3}\right)\left(6,000 \, PSi\right)$
$+ \left(5 \times 10^{-3}\right)\left(44,000 \, PSi\right) =$

$100 \, PSi + 75 \, PSi + 600 \, PSi + 15 \, PSi + 220 \, PSi$

$\simeq 1,010 \, PSi$

Test Yourself 4.1

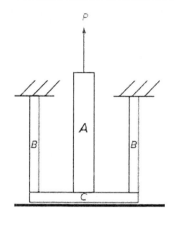

$P = 45 \text{ kN}$

$A: \quad L = 350 \text{ mm}$

$\qquad D = 50 \text{ mm}$

$B: \quad L = 275 \text{ mm}$

$\qquad D = 15 \text{ mm}$

ROD A: MOVES UPWARD $\quad \Delta_A = PL/AE$

$$\Delta_A = (45{,}000 N)(0.350 M) / (\pi(0.050M)^2/4)(70\times10^9 N/M^2)$$

$$= 1.15\times10^{-4} M = 0.115 \text{ mm}$$

ROD B: RODS COMPRESS BUT PLATE C MOVES UPWARD

$$\Delta_B = (45{,}000 N)(0.275 M) / (\tfrac{\pi}{4}(0.015)^2\times 2)(70\times10^9 N/M^2)$$

$$= 5.00\times10^{-4} M \quad OR \quad 0.500 \text{ mm}$$

COMPATIBILITY: $\Delta_A + \Delta_B = \boxed{\Delta_{TOTAL} = 0.115 + 0.500 = 0.615 \text{ mm}}$

Test Yourself 4.2

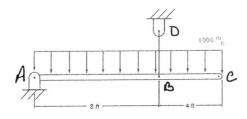

$1,000 \frac{lb}{ft}$

A   B   C

8 ft   4 ft

FBD

$12,000\,lb$ DB

$A_H \rightarrow$

A_V 6'   2'   4'

TOTAL LOAD =

$1000 \frac{lb}{ft}(12ft) = 12,000\,lb$

$\Sigma M_A = 0$

$-(12,000)(6) + DB(8) = 0$

$\therefore DB = 9,000\,lb$

$$\Delta_{DB} = \frac{(9,000\,lb)(4\,ft \cdot 12in/ft)}{\frac{\pi(0.125in)^2}{4}(29\times10^6\,lb/in^2)}$$

$$\boxed{\Delta_{DB} = 1.21\,in}$$

$$\frac{\Delta_C}{12} = \frac{\Delta_{DB}}{8}$$

$$\Delta_C = \frac{12}{8}\Delta_{DB} = 1.5(1.21in)$$

$$\boxed{\Delta_C = 1.82\,in}$$

Test Yourself 4.3

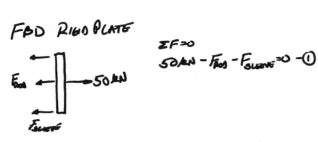

$\Sigma F = 0$

$50 \text{kN} - F_{ROD} - F_{SLEEVE} = 0 \;-①$

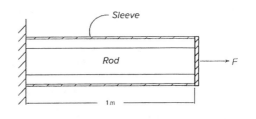

COMPATIBILITY REQUIRES (DUE TO RIGID PLATE)

$\Delta_{ROD} = \Delta_{SLEEVE} \;-②$ $\Delta_{ROD} = \dfrac{F_{ROD}\, L}{A_{ROD}\, E_{ROD}}$

$$\Delta_{SLEEVE} = \dfrac{F_{SLEEVE}\, L}{A_{SLEEVE}\, E_{SLEEVE}}$$

$$\Delta_{ROD} = \dfrac{F_{ROD}\,(1m)}{\frac{\pi (0.075m)^2}{4}\left(200\times10^9\ N/m^2\right)} = 1.132\times10^{-9}\, F_{ROD}$$

$$\Delta_{SLEEVE} = \dfrac{F_{SLEEVE}\,(1m)}{\frac{\pi (0.150^2 - 0.125^2)}{4}\left(70\times10^9\right)} = 2.646\times10^{-9}\, F_{SLEEVE}$$

Sub in these values into Eqn ②

② $1.132\times10^{-9}\, F_{ROD} = 2.646\times10^{-9}\, F_{SLEEVE}$ ⎤ TWO EQNS;

① $50,000 - F_{ROD} - F_{SLEEVE} = 0$ ⎦ TWO UNKS

$\therefore F_{ROD} = \dfrac{2.646\times10^{-9}\, F_{SLEEVE}}{1.132\times10^{-9}} = 2.337\, F_{SLEEVE}$

$50,000 - 2.337\, F_{SLEEVE} - F_{SLEEVE} = 0$

$\therefore \boxed{F_{SLEEVE} = \dfrac{50,000\ N}{3.337} = 14,980\ N}$

$\therefore \boxed{F_{ROD} = 2.337\, F_{SLEEVE} = 35,010\ N}$

Check: $35,010 + 14,980 = 49,990\ N \simeq 50,000\ N$

$$\Delta_{ROD} = \dfrac{F_{ROD}\,(1m)}{\frac{\pi (0.075m)^2}{4}\left(200\times10^9\ N/m^2\right)} = 1.132\times10^{-9}\, F_{ROD}$$

$$= 1.132\times10^{-9}\, F_{ROD} = 1.132\times10^{-9}\,(35,010)$$

$$\boxed{\Delta_{ROD} = 0.0396\ mm}$$

Test Yourself 4.4

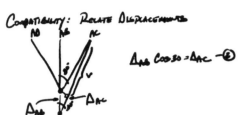

FBD Pɴ A

$\Sigma F_x = 0$
$-AD\sin 30 + AC\sin 30 = 0$
$\therefore AD = AC$

$\Sigma F_y = 0$
$AB + 2AC\cos 30 - 1000 = 0$ —①

Coᴍᴘᴀᴛɪʙɪʟɪᴛʏ: Rᴇʟᴀᴛᴇ Dɪsᴘʟᴀᴄᴇᴍᴇɴᴛs

$\Delta_{AB}\cos 30 = \Delta_{AC}$ —②

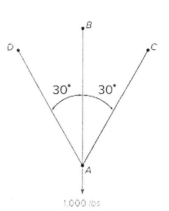

1,000 lbs

USE FL/AE

$$\frac{(AB)L}{AE}\cos 30 = \frac{(AC)L}{AE}$$

$\therefore \quad AB\cos 30 = AC$

SɪɴCᴇ

$AB + 2AC\cos 30 - 1000 = 0$

$AB + 2AB\cos^2 30 - 1000 = 0$

$AB(1 + 2\cos^2 30) = 1000$

$\therefore \boxed{AB = 400\ \text{lb}}$

$AB\cos 30 = \boxed{\begin{aligned}AC &= 346.4\ \text{lb}\\ AD &= 346.4\ \text{lb}\end{aligned}}$

Since $d = \tfrac{1}{4}"$ $A = \dfrac{\pi(0.25in)^2}{4} = 0.04909\ in^2$

Nᴏʀᴍᴀʟ Sᴛʀᴇssᴇs:

$$\boxed{\sigma_{AB} = \frac{400\ lb}{0.04909\ in^2} = 8,148\ psi}$$

$$\boxed{\sigma_{AC} = \sigma_{AD} = \frac{346.4}{0.04909} = 7,056\ psi}$$

$\Delta = \Delta_{AB} = \dfrac{(AB)L}{AE}$

$\Delta = \dfrac{(400\ lb)(5ft \times 12in/ft)}{(0.04909\ in^2)(29\times 10^6\ lb/in^2)}$

$$\boxed{\Delta = 0.0169\ in}$$

Test Yourself 4.5

FBD

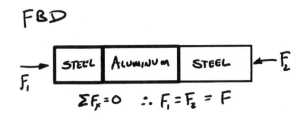

$\Sigma F_x = 0 \quad \therefore F_1 = F_2 = F$

COMPATIBILITY $\quad \Sigma(\Delta_A + \Delta_B + \Delta_C) = 0$
DUE TO RIGID WALLS

CONSTITUTIVE LAW: $\quad \Delta = \frac{FL}{AE} + \alpha L \Delta T$

(A) $\dfrac{F(3_{in})}{(1_{in^2})(29 \times 10^6 \, lb/in^2)} + (6 \times 10^{-6}/°F)(3_{in})(-70°F)$

$\quad 10\ 3.45 \times 10^{-9} F - 1.2600 \times 10^{-3}$

(B) $\dfrac{F(6_{in})}{(1_{in^2})(10 \times 10^6 \, lb/in^2)} + (12 \times 10^{-6}/°F)(6_{in})(-70°F)$

$\quad 600.00 \times 10^{-9} F - 5.0400 \times 10^{-3}$

(C) $\dfrac{F(6_{in})}{(1_{in^2})(29 \times 10^6 \, lb/in^2)} + (6 \times 10^{-6}/°F)(6_{in})(-70°F)$

$\quad 206.90 \times 10^{-9} F - 2.5200 \times 10^{-3}$

(A) + (B) + (C) = 0

$\left(10\ 3.45 \times 10^{-9} F - 1.2600 \times 10^{-3}\right)$
$+ \left(600.00 \times 10^{-9} F - 5.0400 \times 10^{-3}\right)$
$\quad + \left(206.90 \times 10^{-9} F - 2.5200 \times 10^{-3}\right) = 0$

$\quad 910.35 \times 10^{-9} F - 8.82 \times 10^{-3} = 0$

$\therefore \boxed{F = 9688.5 \, lb}$

$\therefore \boxed{\sigma_{normal} = F/A = 9688.5 \, psi}$

Test Yourself 4.6

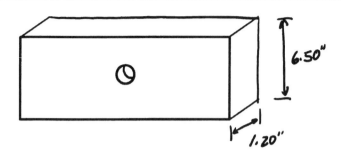

6.50"

1.20"

$r = 1/8" = 0.125"$
$w = 6.50"$
$t = 1.20"$
$N = 10,000\, lb$

$$\sigma_{AVG} = \frac{10,000\, lb}{[6.50 - 2(0.125)]\,1.20} = 1,333\, psi$$

$Chart: \quad \frac{2r}{w} = \frac{2(0.125")}{6.50"} = 0.038 \quad \therefore K \approx 2.83$

$$\sigma = (1,333\, psi)\, 2.83 = 3,772\, psi$$

Test Yourself 5.1

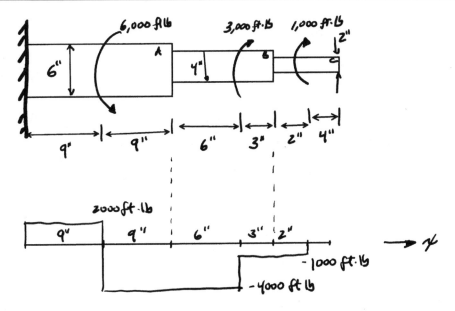

USE THE MAXIMUM TORQUE IN EACH SECTION.

SHAFT A

$r = 3''$ $J = \frac{1}{2}\pi (3)^4 = 127.23\,in^4$

$\tau_{MAX} = \left(\frac{Tc}{J}\right)_A = \dfrac{(-4000\,ft\cdot lb)(12\,in/ft)(3\,in)}{127.23\,in^4} = 1,131.8\,psi$

SHAFT B

$r = 2''$ $J = \frac{1}{2}\pi (2)^4 = 25.133\,in^4$

$\tau_{MAX} = \left(\frac{Tc}{J}\right)_B = \dfrac{(-4000\,ft\cdot lb)(12\,in/ft)(2\,in)}{25.133\,in^4} = 3,819.7\,psi$

SHAFT C

$r = 1''$ $J = \frac{1}{2}\pi (1)^4 = 1.5708\,in^4$

$\tau_{MAX} = \left(\frac{Tc}{J}\right)_C = \dfrac{(-1000\,ft\cdot lb)(12\,in/ft)(1\,in)}{1.5708\,in^4} = 7,639.4\,psi$

Test Yourself 5.2

$\boxed{D_1}$

$P = 7{,}000 \, kW = 7 \times 10^6 \, NM/s$ $\omega_1 = 5{,}500 \, \frac{Rev}{MIN} \cdot \frac{1 \, min}{60 \, s} \cdot \frac{2\pi \, rad}{1 \, Rev}$

$\therefore T = P/\omega_1$ $\omega_1 = 575.96 \, rad/s$

$= \dfrac{7 \times 10^6 \, NM/s}{575.96 \, rad/s} = 12{,}154 \, N \cdot M$

$\tau_{Allow} = 100 \, MPa = Tc/J \Rightarrow \boxed{\begin{array}{c} c = r_1 \\ J = \frac{\pi r_1^4}{2} \end{array}}$

$100 \times 10^6 \, \dfrac{N}{M^2} = \dfrac{(12{,}154 \, NM) \, r_{1 \, \pi}}{\frac{1}{2} \pi r_1^{\cancel{4}3}}$

$\frac{1}{2} \pi r_1^3 = \dfrac{(12{,}154) \, M}{100 \times 10^6 / M^2} = 121.5 \times 10^{-6} \, M^3$

$r_1^3 = \dfrac{121.5 \times 10^{-6} \, M^3}{\frac{1}{2} \pi}$ $\therefore r_1 = 0.0426 \, M = 42.6 \, mm$

$\boxed{\therefore D_1 = 85.2 \, mm}$

$\boxed{D_2}$ $P = 7 \times 10^6 \, \dfrac{NM}{s}$ $\omega_2 = \dfrac{\omega_1}{10} = \dfrac{575.96}{10}$

$\therefore \omega_2 = 57.596 \, radians/s$

$T = P/\omega = 7 \times 10^6 / 57.596 = 1.215 \times 10^5 \, N \cdot M$

$\tau_{All} = 100 \times 10^6 \, \dfrac{N}{M^2} = \dfrac{T r_2}{J} = \dfrac{(1.215 \times 10^5)(r_2)}{\frac{1}{2} \pi r_2^4}$

$\frac{1}{2} \pi r_2^3 = \dfrac{1.215 \times 10^5}{100 \times 10^6}$

$r_2 = 0.0918 \, M$ $\therefore \boxed{D_2 = 183.6 \, mm}$

$r_2 = 91.8 \, mm$

Test Yourself 5.3

$$2° \cdot \frac{2\pi\, rad}{360°} = 0.0349\, rad$$

$$J = \tfrac{1}{2}\pi r^4 = \tfrac{1}{2}\pi (5in)^4 = 981.7 in^4$$

$$L = 8.5 ft * 12 = 102\, in$$

$$\phi = TL/GJ$$

$$0.0349 = (T)(102) \big/ (15\times10^6)(981.7)$$

$$\therefore T = 5.038 \times 10^6\ in\text{-}lb$$

$$\omega = (150\, rpm)(2\pi/60) = 15.71\ rad/s$$

$$P = T\cdot\omega = (5.038\times10^6\ in\text{-}lb)(15.71\ rad/s)$$
$$= 79.15\times10^6\ in\text{-}lb/s$$

$$P = \left(79.15\times10^6\ \frac{in\cdot lb}{s}\right)\left(\frac{1ft}{12in}\right)\left(\frac{1hp}{550ft\cdot lb/s}\right)$$

$$\boxed{P = 11,992\ hp}$$

$$\tau = \frac{Tc}{J}$$

$$= \frac{(5.038\times10^6\ in\cdot lb)(5in)}{\tfrac{1}{2}\pi(5)^4}$$

$$\boxed{\tau = 25,658\ psi}$$

Test Yourself 5.4

$$\omega_A = 500 \frac{Rev}{MIN} \cdot \frac{1 MIN}{60S} \cdot \frac{2\pi \, radians}{1 Rev}$$

$$\omega_A = 52.36 \text{ radians}/s$$

T_B $\therefore \omega_B = 2\omega_A$

$$\tau_A = \frac{(175 N \cdot m)(0.035/2)m}{\frac{1}{2}\pi(0.035/2)^4 m^4} = 20.79 \, mPa$$

WE SET THIS TO τ_B

$$P = T_A \cdot \omega_A = (175 N \cdot m)(52.36 \text{ radians}/s)$$

$$\therefore P = 9,162.98 \text{ WATTS}$$

FOR B

$$T_B = \frac{P}{\omega_B} = \frac{9162.98}{2\omega_A} = \frac{9162.98}{2(52.36)} = 87.5 N \cdot m$$

$$\tau_B = 20.79 \times 10^6 \frac{N}{M^2} = \frac{T_B \, r_B}{J_B}$$

$$20.79 \times 10^6 \frac{N}{M^2} = \frac{(87.5 N \cdot m)(r_B)m}{\frac{1}{2}\pi(r_B)^4 \, m^4}$$

$$20.79 \times 10^6 = \frac{55.7}{r_B^3} = 13.9 \times 10^{-3} m$$

$$\boxed{\therefore \; r_B = 13.9 mm \; (27.8 mm \, diameter)}$$

Test Yourself 5.5

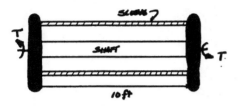

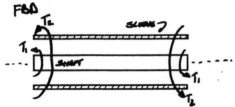

T_2 : TORQUE IN SLEEVE

T_1 : TORQUE IN SHAFT

$\Sigma M = T_1 + T_2 = T = 5000 \text{ ft·lb}$ —①

$J_{SHAFT} = \frac{1}{2}\pi(2.5)^4 = 61.36 \text{ in}^4$

$J_{SLEEVE} = \frac{1}{2}\pi(7.5^4 - 6^4) = 2934 \text{ in}^4$

COMPATIBILITY REQUIRES: $\phi_{SHAFT} = \phi_{SLEEVE}$

$$\frac{T_1 L}{(11\times10^6)(61.36)} = \frac{T_2 L}{(33\times10^6)(2934)} \quad —②$$

$\therefore 1.482\times10^{-9}T_1 = 8.964\times10^{-5}T_2$

OR $T_2 = 16.52 T_1$

FROM ① $T_1 + T_2 = 5000(12) = 60,000 \text{ in·lb}$

$T_1 + 16.52T_1 = 60,000$

$\therefore \boxed{\begin{array}{l} T_1 = 3,425 \text{ in·lb} \\ T_2 = 56,581 \text{ in·lb} \end{array}}$

$\boxed{\begin{array}{l} \tau_{SHAFT} = \dfrac{(3425 \text{ in·lb})(2.5 \text{ in})}{61.36 \text{ in}^4} = 139.5 \text{ psi} \\[2mm] \tau_{SLEEVE} = \dfrac{(56,581 \text{ in·lb})(7.5 \text{ in})}{2934 \text{ in}^4} = 144.6 \text{ psi} \end{array}}$

Test Yourself 5.6

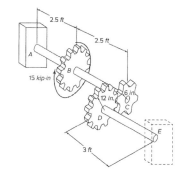

Since Gear D is twice the size of Gear C, we have: $\boxed{2\phi_D = \phi_C}$ —①

SHAFT ABC

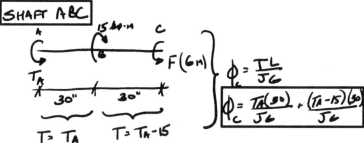

$T = T_A$ $T = T_A - 15$

$\phi_C = \dfrac{TL}{JG}$

$\boxed{\phi_C = \dfrac{T_A(30)}{JG} + \dfrac{(T_A - 15)(30)}{JG}}$

SHAFT DE

$\therefore T = T_E$

$\phi_D = \dfrac{TL}{JG}$

$\boxed{\phi_D = \dfrac{T_E \, 36}{JG}}$

From ① $\boxed{2\phi_D = \phi_C}$ Note:

$\therefore 2\left[\dfrac{T_E(\cancel{36})}{\cancel{JG}}\right] = \left[\dfrac{T_A(30)}{\cancel{JG}} + \dfrac{(T_A - 15)(30)}{\cancel{JG}}\right]$ cancel terms

$72 T_E = 30 T_A + 30 T_A - 450$

$\boxed{72 T_E = 60 T_A - 450}$ —②

FBD - System

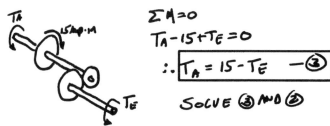

$\Sigma M = 0$

$T_A - 15 + T_E = 0$

$\therefore \boxed{T_A = 15 - T_E}$ —③

Solve ③ and ②

$72 T_E = 60(15 - T_E) - 450$

$132 T_E = 450 \quad \therefore \; T_E = 3.409 \text{ kip·in}$

$\therefore T_A = 15 - T_E \quad T_A = 11.59 \text{ kip·in}$

$\boxed{\begin{array}{l} T_E = 5.14 \text{ kip·in} \\ T_A = 9.886 \text{ kip·in} \end{array}}$

ON SHAFT A-B-C, THE
TORQUE ON SECTION $\overline{AB}$ IS
GIVEN AS

9.866 Kip·in   15 Kip·in

$\therefore T = (9.866 - 15) = -5.134$ Kip·in

$\phi = \dfrac{TL}{JG}$

$L = 2.5' = 30$ in

$T = -5.134$ Kip·in
$\quad = -5134$ in·lb

$G = 11,300,000$ psi

$J = \dfrac{\pi c^4}{2} = \dfrac{\pi}{2}(2)^4 = 25.133 \text{ in}^4$

$\phi = \dfrac{(-5134 \text{ in·lb})(30 \text{ in})}{(11.3 \times 10^6 \text{ lb/in}^2)(25.133 \text{ in}^4)} = \boxed{5.423 \times 10^{-4} \text{ radians}}$

$(5.423 \times 10^{-4} \text{ radians}) \dfrac{360 \text{ degrees}}{2\pi \text{ radians}}$

$\boxed{\phi = 0.0311°}$

Test Yourself 6.1

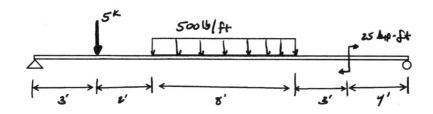

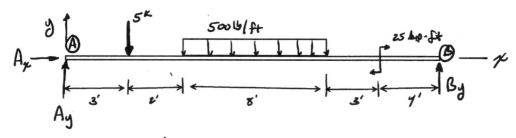

$$\sum F_x = 0 \quad \therefore A_x = 0$$

$$\sum M_A = 0 \quad -(5000)(3) - (500)(8)(9) - 25000 + B_y(20) = 0$$

$$\therefore B_y = 3,800 \text{ lb (upward)}$$

$$\sum F_y = 0 \quad A_y - 5000 - 500(8) + 3,800 = 0$$

$$\therefore A_y = 5,200 \text{ lb (upward)}$$

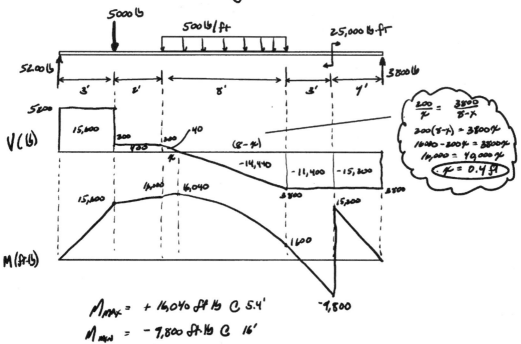

$$\frac{200}{x} = \frac{3800}{8-x}$$

$$200(8-x) = 3800x$$

$$1600 - 200x = 3800x$$

$$16,000 = 40,000x$$

$$x = 0.4 \text{ ft}$$

$$M_{MAX} = +16,040 \text{ ft-lb} \ @ \ 5.4'$$

$$M_{MIN} = -7,800 \text{ ft-lb} \ @ \ 16'$$

Test Yourself 6.2

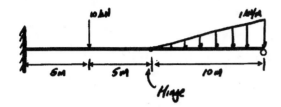

FBD - BC

$\Sigma F_x = 0$ ∴ $\boxed{B_H = 0}$

$\Sigma M_B = 0$ $C_v(10) - \left[\frac{1}{2}(10)(1)\right]\frac{2}{3}(10) = 0$

∴ $\boxed{C_v = 3.333 \text{ kN}}$

$\Sigma F_y = 0$ $B_v + 3.333 - \frac{1}{2}(10)(1) = 0$

∴ $\boxed{B_v = 1.667 \text{ kN}}$

NOTE: Let's compute where the shear
equals <u>zero</u> in this segment.
We need this to find the
location AND MAGNITUDE
of The Maximum Moment.

FBD CUT AT Position "x"

$\Sigma F_y = 0$ $1.667 - \frac{1}{2}(0.1x)(x) - V = 0$

$V = 1.667 - 0.05 x^2$

Set $V = 0$ ∴ $x = 5.77 \text{ m}$
⟹ Location of M_{MAX}

FOR THE MAGNITUDE, COMPUTE THE
AREA UNDER THE SHEAR DIAGRAM

$$M_{MAX} = \int_{0}^{5.77} (1.667 - 0.05x^2)\,dx$$

$$= \left[1.667x - \frac{0.05}{3}x^3\right]\Bigg|_{0}^{5.77}$$

$$\Rightarrow M_{MAX} = 6.42 \text{ kN·m}$$

FBD - AB

10 kN

M_A

A_H

A_v

5m   5m

1.667 kN

$\Sigma F_x = 0$ ∴ $\boxed{A_H = 0}$

$\Sigma F_y = 0$ $-1.667 - 10 + A_v = 0$

∴ $\boxed{A_v = 11.667 \text{ kN}}$

NOTE: B_v HAS AN EQUAL MAGNITUDE
BUT OPPOSITE DIRECTION!

$\Sigma M_A = 0$

$-M_A - 10(5) - 1.667(10) = 0$

∴ $\boxed{M_A = -66.667 \text{ kN·m}}$

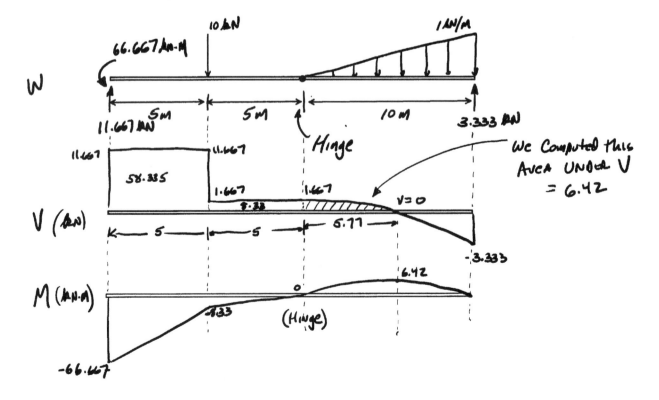

W

66.667 kN·M

10 kN

1 kN/M

5m   5m   10m

11.667 kN   Hinge   3.333 kN

V (kN)

11.667

58.335

11.667

1.667

8.33

1.667

V = 0

We Computed this
Area under V
= 6.42

5   5   5.77

-3.333

M (kN·M)

6.42

-8.33

(Hinge)

-66.667

Test Yourself 6.3

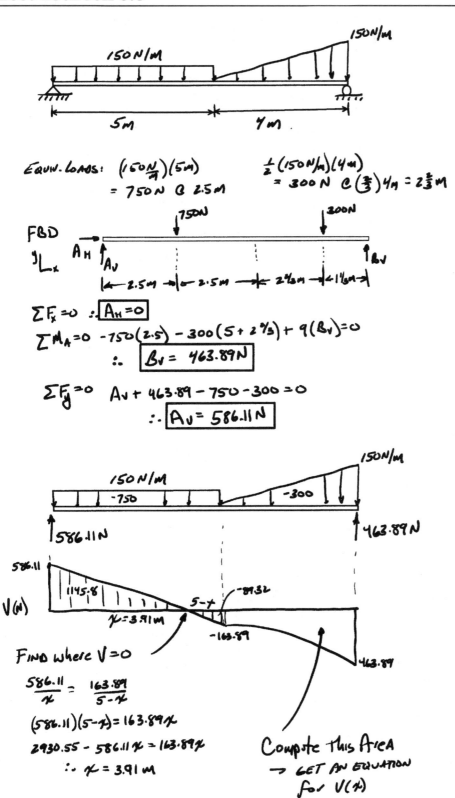

EQUIV. LOADS: $\left(150\frac{N}{m}\right)(5m)$ $\frac{1}{2}(150N/m)(4m)$
                = 750 N @ 2.5 M   = 300 N @ $\left(\frac{2}{3}\right)4m = 2\frac{2}{3}m$

FBD

$\sum F_x = 0$ ∴ $\boxed{A_H = 0}$

$\sum M_A = 0$ $-750(2.5) - 300(5 + 2\frac{2}{3}) + 9(B_v) = 0$
                ∴ $\boxed{B_v = 463.89N}$

$\sum F_y = 0$ $A_v + 463.89 - 750 - 300 = 0$
                ∴ $\boxed{A_v = 586.11N}$

FIND WHERE V=0

$\frac{586.11}{x} = \frac{163.89}{5-x}$

$(586.11)(5-x) = 163.89x$

$2930.55 - 586.11x = 163.89x$

∴ $x = 3.91m$

COMPUTE THIS AREA
→ GET AN EQUATION
   for V(x)

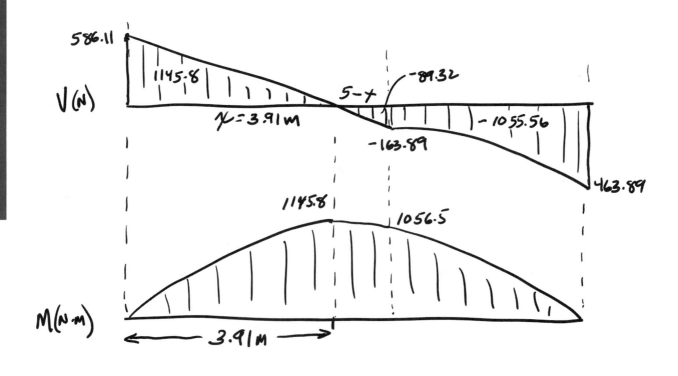

$$\Sigma F_y = 0 \quad 586.11 - 750 - \tfrac{1}{2}\left(\tfrac{150}{4}\right)(x)x = 0$$

$$V(x) = 586.11 - 750 - 18.75x^2$$

$$V(x) = -163.89 - 18.75x^2$$

$$Area = \int_0^4 (-163.89 - 18.75x^2)\,dx = \left[-163.89x - \tfrac{18.75}{3}x^3\right]_0^4$$

$$Area = -1055.6 \quad N\cdot m$$

586.11

V(N)

1145.8   x=3.91m   5-x   -89.32   -163.89   -1055.56   463.89

M(N·m)   1145.8   1056.5   3.91M

Test Yourself 6.4

(1) Find the Centroid

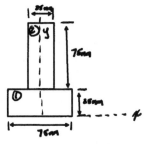

Section	Area	$\bar{y}_i$	$A_i\bar{y}_i$
1	1875	12.5	23,437.5
2	1875	62.5	117,187.5
Σ	3750		140,625

$$\therefore \boxed{\bar{y} = {}^{140,625}/_{3750} = 37.5 \text{ mm}}$$

(2) Compute I_{xx} About the Centroid

$$I_{x_c x_c} = \frac{1}{12}(75)(25)^3 + (75)(25)(37.5 - 12.5)^2$$

$$+ \frac{1}{12}(25)(75)^3 + (75)(25)\left((25 + 75/2) - 37.5\right)^2$$

$$= 97,656.25 + 1171875.$$

$$+ 878,906.25 + 1171875$$

$$I_{x_c x_c} = 3,320,312.5 \text{ MM}^4$$

$$\boxed{I_{x_c x_c} = 3.32 \times 10^{-6} \text{ M}^4}$$

$$\sigma_{zz} = -\frac{Mc}{I_{x_c x_c}}$$

$$= -\frac{(4,500 \text{ NM})(0.0625 \text{ m})}{3.32 \times 10^{-6} \text{ M}^4} = -84.7 \times 10^6 \text{ N/M}^2$$

$$\boxed{\sigma_{Top} = 84.7 \text{ MPa Compression}}$$

$$= -\frac{(4,500 \text{ NM})(-0.0375 \text{ m})}{3.32 \times 10^6 \text{ M}^4} = 50.8 \times 10^6 \text{ N/M}^2$$

$$\boxed{\sigma_{Bottom} = 50.8 \text{ MPa Tension}}$$

Test Yourself 6.5

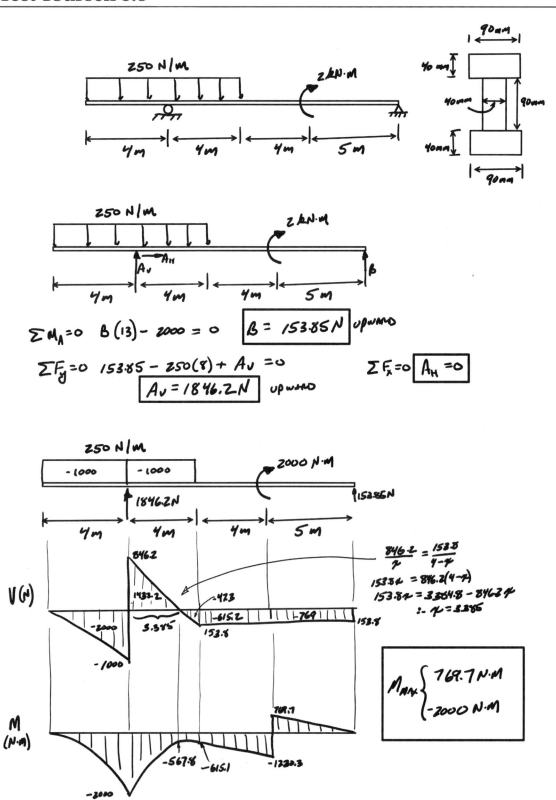

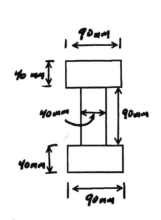

FROM SYMMETRY, THE CENTROID IS AT THE center of the X-SEC.

$$I_{zz} = \frac{1}{12}(90)(40)^3 + (90)(40)(65)^2 +$$
$$\frac{1}{12}(40)(90)^3 +$$
$$\frac{1}{12}(90)(40)^3 + (90)(40)(65)^2$$

$$I_{zz} = 33,810,000 \text{ mm}^4 \frac{(0.001\text{m})^4}{(1\text{mm})^4}$$
$$= 3.381 \times 10^{-5} \text{ m}^4$$

Compute the bending stress:

$$\sigma_{TEN} = \frac{(2000 \text{ N·m})(0.085\text{m})}{3.381 \times 10^{-5} \text{ m}^4} = 5.03 \text{ MPa}$$

$$\sigma_{COMP} = \frac{(-2000 \text{ N·m})(0.085\text{m})}{3.381 \times 10^{-5} \text{ m}^2} = -5.03 \text{ MPa}$$

Test Yourself 6.6

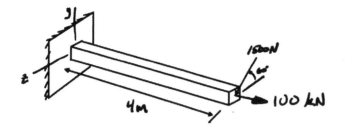

① The loading will produce
At the base of the beam:

$F_x = 100\ kN$

$M_z = -[(1500N)\sin 60]\ 4m = -5196\ N\cdot M$

$M_y = -(1500\ N)(\cos 60)]\ 4m = -3000\ N\cdot M$

② May stress on x-section
Due to F_x

$\sigma_x = \dfrac{100,000N}{(0.100m)^2} = 10\ MPa$

③ May stress on x-section
Due to M_z

$\sigma_x = \left|\dfrac{M_z c}{I}\right| = \left|\dfrac{(-5196\ N\cdot M)(0.050M)}{\frac{1}{12}(0.100)^4}\right|$

$= 31.2\ MPa$

④

MAX STRESS ON X-SECTION
DUE TO M_y

$$\sigma_x = \left|\frac{M_q c}{I}\right| = \left|\frac{(-3000\,Nm)(0.050m)}{\frac{1}{12}(0.100)^4}\right|$$

$$= 18.0\ MPa$$

$-18MPa$ C   18MPa
T

M_y

$-18MPa$ C   T 18MPa

⑤ SUPERPOSE the 3 CASES:

M_z F M_y
M_z

23.2 MPa (T)   59.2 MPa (T)

31.2 MPa (C)   3.20 MPa (C)

Test Yourself 6.7

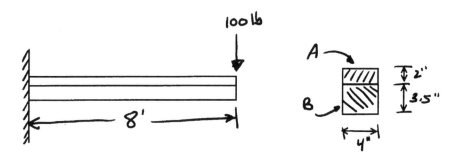

① Build the equivalent section
→ USE MATERIAL A AS "BASE"

∴ WIDTH of B IS EXPANDED BY

$$E_B/E_A = 17.5 \times 10^6 / 7 \times 10^6 = 2.5$$

$$(4_{in}) 2.5 = 10 in$$

② Compute Centroid ; I

Section	Area;	$\bar{y}_i$	$A_i \bar{y}_i$
A	$8 in^2$	$4.5 in$	$36 in^3$
B	$35 in^2$	$1.75 in$	$61.25 in^3$
	$43 in^2$		$97.25 in^3$

$$\boxed{\bar{y} = 97.25/43 = 2.26 in}$$

$$I = \tfrac{1}{12}(4)(2)^3 + (4)(2)(4.5 - 2.26)^2$$
$$+ \tfrac{1}{12}(10)(3.5)^3 + (10)(3.5)(2.26 - 1.75)^2$$

$$\boxed{I_{eq} = 87.64 \text{ in}^4}$$

③ LOADING: The Maximum Bending Moment is at the Fixed End of the Beam

$$\therefore M_{max} = -(8 ft)(100 lb) = -800 \text{ ft lb}$$

④ Compute Stresses $\sigma = -Mc/I$

$$\sigma_A = -\frac{(-800 \text{ ft lb})(12 \text{ in}/ft)(5.5 \text{ in} - 2.26 \text{ in})}{87.64 \text{ in}^4} = 354.9 \text{ psi (T)}$$

$$\sigma_{B_{eqv}} = -\frac{(-800)(12)(-2.26)}{87.64} = -247.6 \text{ psi (C)}$$

↳ Must Convert Material B back to the Original Material:

$$E_B/E_A = 2.5 \qquad \therefore \sigma_B = 2.5(-247.6) = -619.0 \text{ psi (C)}$$

Bending Stress: 354.9 psi Tension (Top)
619.0 psi Compression (Bottom)

Test Yourself 7.1

Compute the maximum Q for the shape shown. The wall thickness t, is 10mm on all sides.

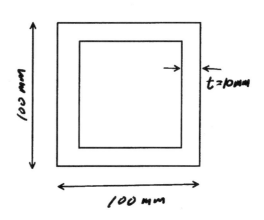

For this shape, Q_{max} will occur at the centroid

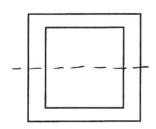 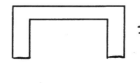 =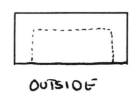

OUTSIDE   INSIDE

	A_i	$\bar{y}_i$	$A_i \bar{y}_i$
OUTSIDE	$100mm \times 50mm = 5,000\ mm^2$	25mm	$125,000\ mm^3$
INSIDE	$-80mm \times 40mm = -3,200\ mm^2$	20mm	$-64,000\ mm^3$

$$Q = \Sigma = 61,000\ mm^3$$

$$\boxed{Q = 61,000\ mm^3}$$

WE CAN Also Compute this A Different Way:

	$\bar{y}_i\ (mm)$	$A_i\ (mm^2)$	$A_i \bar{y}_i\ (mm^3)$
①	25	$(50)(10) = 500$	12,500
②	45	$(80)(10) = 800$	36,000
③	25	$(50)(10) = 500$	12,500

$$\boxed{Q = 61,000\ mm^3}\ \checkmark$$

$t = 10mm$

Test Yourself 7.2

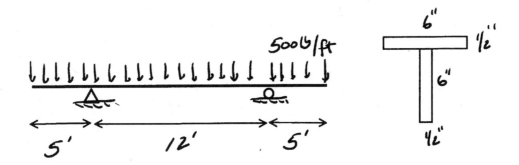

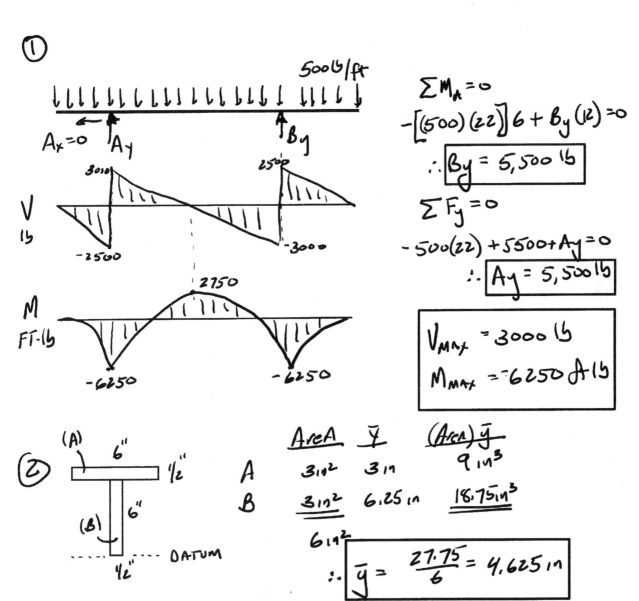

$\Sigma M_A = 0$

$-[(500)(22)]6 + B_y (12) = 0$

$\therefore \boxed{B_y = 5,500 \text{ lb}}$

$\Sigma F_y = 0$

$-500(22) + 5500 + A_y = 0$

$\therefore \boxed{A_y = 5,500 \text{ lb}}$

$$\boxed{\begin{array}{l} V_{MAX} = 3000 \text{ lb} \\ M_{MAX} = -6250 \text{ ft lb} \end{array}}$$

	Area	$\bar{y}$	(Area)$\bar{y}$
A	3 in²	3 in	9 in³
B	3 in²	6.25 in	18.75 in³
	6 in²		

$\therefore \boxed{\bar{y} = \dfrac{27.75}{6} = 4.625 \text{ in}}$

$$I_c = \tfrac{1}{12}(\tfrac{1}{2})(6)^3 + (3)(3-4.625)^2$$
$$+ \tfrac{1}{12}(6)(\tfrac{1}{2})^3 + (3)(1.625)^2$$
$$= [9+7.922] + [0.0625+7.922]$$
$$\boxed{I_c = 24.906 \text{ in}^4}$$

$$\therefore \quad \boxed{\sigma_{xx} = -\frac{(-6250)(12)(-4.625)}{24.906} = -13{,}927 \, psi}$$

MAX SHEAR STRESS AT NEUTRAL AXIS

$$\tau_{xy} = \left|\frac{VQ}{Ib}\right| \qquad V_{MAX} = 3000 \, lb \quad I = 24.906 \text{ in}^4$$
$$b = \tfrac{1}{2}"$$

$$|Q| = (\tfrac{1}{2}in)(4.625 in)(4.625 in/2)$$
$$= 5.348 \text{ in}^3$$

$$\boxed{|\tau_{xy}| = \frac{(3000)(5.348)}{(24.906)(\tfrac{1}{2})} = 1{,}288 \, psi}$$

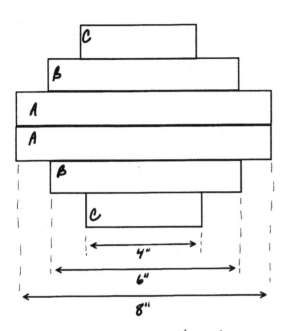

C
B
A
A
B
C

4"
6"
8"

$(4)(1)(2.5) = 10$ $t = 4$ $10/4 = 2.5$
$(4)(1)(2.5) + (6)(1)(1.5) = 19$ $t = 6$ $19/6 = 3.17$
$10 + 9 + (8)(1)(0.5) = 23$ $t = 8$ $23/8 = 2.875$

$\tau = \dfrac{VQ}{It}$ FOR A GIVEN SECTION V AND I ARE CONSTANT.

$\therefore$ INVESTIGATE Q/t

INTERFACE A-A
$Q = (8)(1)(0.5) + (6)(1)(1.5) + (4)(1)(2.5) = 23 \, in^3$
$t = 8 \, in$ $\therefore Q/t = 23/8 = 2.875 \, in^2$

INTERFACE A-B
$Q = (6)(1)(1.5) + (4)(1)(2.5) = 19 \, in^3$
$t = 6 \, in$ $\therefore Q/t = 19/6 = 3.167 \, in^2$

INTERFACE B-C
$Q = (4)(1)(2.5) = 10 \, in^3$
$t = 4 \, in$ $Q/t = 10/4 = 2.500 \, in^2$

$\therefore$ | HIGHEST STRESS AT A-B INTERFACE |

Test Yourself 7.3

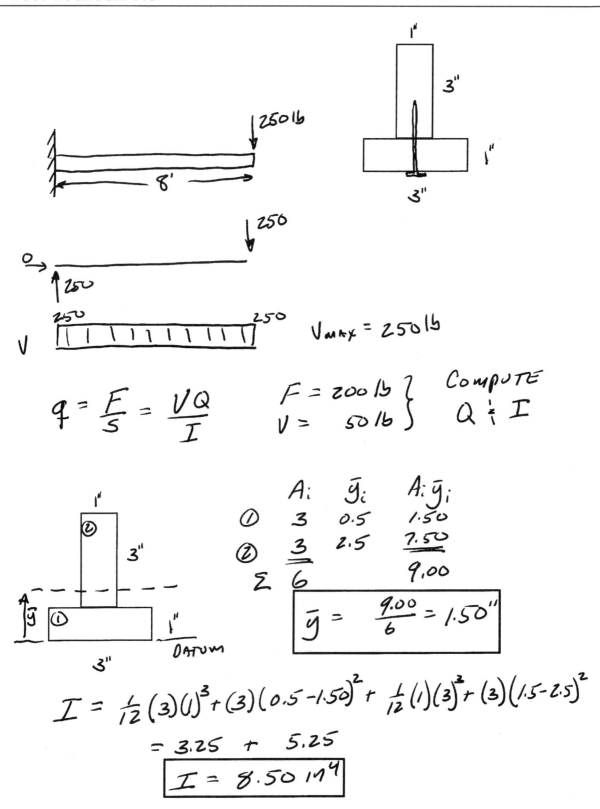

$$q = \frac{F}{S} = \frac{VQ}{I}$$

$$\left. \begin{matrix} F = 200\,lb \\ V = 50\,lb \end{matrix} \right\} \quad \text{COMPUTE } Q \text{ \& } I$$

	A_i	$\bar{y}_i$	$A_i\bar{y}_i$
①	3	0.5	1.50
②	3	2.5	7.50
Σ	6		9.00

$$\bar{y} = \frac{9.00}{6} = 1.50''$$

$$I = \frac{1}{12}(3)(1)^3 + (3)(0.5-1.50)^2 + \frac{1}{12}(1)(3)^3 + (3)(1.5-2.5)^2$$

$$= 3.25 + 5.25$$

$$\boxed{I = 8.50\ in^4}$$

$V_{MAX} = 250\,lb$

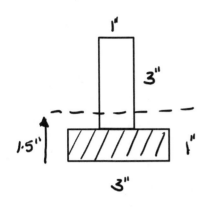

Compute Q AT THE JUNCTION
OF THE TWO PIECES

$$Q = (3")(1") \cdot (1.5" - 0.5")$$

$$\boxed{Q = 3 \text{ in}^3}$$

$$\frac{F}{S} = \frac{VQ}{I} \Rightarrow \frac{200 \text{ lb}}{S \text{ in}} = \frac{(250 \text{ lb})(3 \text{ in}^3)}{8.50 \text{ in}^4}$$

$$\frac{200 \text{ lb}}{S \text{ in}} = 88.24 \text{ lb/in}$$

$$S = 2.27"$$

USE $\boxed{2.25"}$

ROUND DOWN
FOR SAFER
SPACING AND
EASIER TO
MEASURE

Test Yourself 8.1

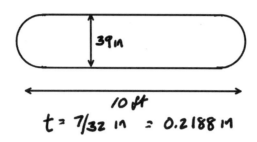

$$t = 7/32 \text{ in} = 0.2188 \text{ in}$$

① $\sigma_{Hoop} = \dfrac{pr}{t}$

$$= \dfrac{(250 \text{ lb/in}^2)(39/2 - 0.2188) \text{ in}}{0.2188 \text{ in}}$$

$$\boxed{\sigma_{Hoop} = 22{,}031 \text{ psi}}$$

② Limit $\sigma_{Hoop} = 20$ ksi, solve for p

$$20{,}000 \text{ psi} = \dfrac{p\,(39/2 - 0.2188) \text{ in}}{0.2188 \text{ in}}$$

$$\boxed{p_{Max} = 227 \text{ psi}}$$

Test Yourself 8.2

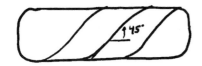

CONSIDER A UNIT SLICE AT THE SEAM

$$\sum F_x = 0 = \left(\frac{pr}{2t}\right)(t)(1) - N\left(t\sqrt{2}\right)\cos 45 + T\left(t\sqrt{2}\right)\cos 45$$

$$\sum F_y = 0 = \left(\frac{pr}{t}\right)(t)(1) + N\left(t\sqrt{2}\right)\sin 45 + T\left(t\sqrt{2}\right)\sin 45$$

SOLVE FOR N ; T

① $\frac{1}{2}pr - Nt + Tt = 0$

② $-pr + Nt + Tt = 0$

ADD ① ; ② $+$ $-\frac{pr}{2} + 0 + 2Tt = 0$

$$\therefore \boxed{T = \frac{1}{4}\frac{pr}{t}}$$

SUB BACK INTO ①

$$\frac{pr}{2} - Nt + \frac{pr}{4t}(t) = 0$$

$$\frac{pr}{2} - Nt + \frac{pr}{4} = 0$$

$$\frac{3}{4}pr = Nt$$

$$\therefore \boxed{N = \frac{3}{4}\frac{pr}{t}}$$

AT 0° $\begin{cases} \text{Hoop:} & \frac{pr}{t} = \frac{pr}{t} \\ \text{Long.:} & \frac{pr}{2t} = \frac{1}{2}\frac{pr}{t} \end{cases}$

AT 45° $\begin{cases} N = \frac{3}{4}\frac{pr}{t} \\ T = \frac{1}{4}\frac{pr}{t} \end{cases}$

Test Yourself 8.3

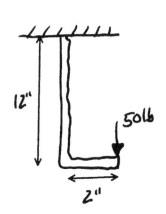

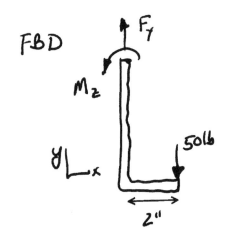

FBD

$\Sigma F_y = 0$

$F_y - 50 = 0$

$\therefore \boxed{F_y = 50 \text{ lb}}$

$\Sigma M_z = 0$

$M_z - (50 \text{ lb})(2 \text{ in}) = 0$

$\therefore \boxed{M_z = 100 \text{ in lb}}$

① AXIAL DUE TO F_y

$$\sigma_{AXIAL} = F_y/A = \frac{50 \text{ lb}}{\pi (0.25 \text{ in})^2/4} = 1,019 \text{ psi (Tension)}$$

② Bending Due to M_z

$$\sigma_{bend} = \pm \frac{M_z C}{I} = \pm \frac{(100 \text{ in·lb})(0.25 \text{ in})/2}{\pi (0.25)^4/64}$$

$$= \pm 65,190 \text{ psi}$$

$\therefore$
$$\boxed{\begin{array}{l} \text{MAX Tension} = 65,190 + 1,019 = 66,210 \text{ psi} \\ \text{MAX Compression} = -65,190 + 1,019 = 64,170 \text{ psi} \end{array}}$$

Test Yourself 8.4

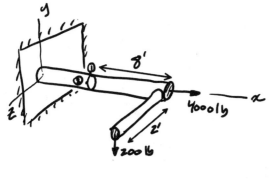

FBD

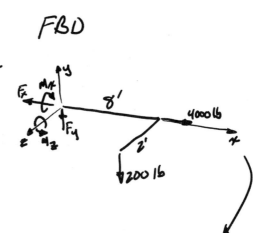

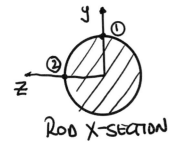

ROD X-SECTION

REACTIONS:

$$\Sigma F_x = 0 \quad \therefore \quad F_x = 4000\,lb$$

$$\Sigma F_y = 0 \quad \quad F_y = 200\,lb$$

$$\Sigma M_x = 0 \quad -M_x + (200\,lb)(2ft) = 0$$
$$\therefore M_x = 400\,ftlb$$
$$= 400 \times 12 = 4,800\,in\,lb$$

$$\Sigma M_z = 0$$
$$M_z - 200\,lb(8ft) = 0$$
$$\therefore M_z = 1600\,ftlb$$
$$= 1600(12) = 19,200\,in\,lb$$

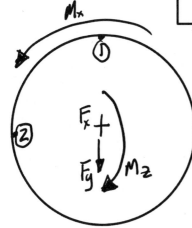

4 LOADS ON THE CROSS-SECTION
AT 8'

$$F_x = 4,000\,lb \quad M_x = 4,800\,in\,lb$$
$$F_y = -200\,lb \quad M_z = -19,200\,in\,lb$$

NOTE: THIS IS THE OUTWARD
FACE AT 8 ft

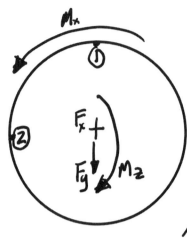

4 LOADS ON THE CROSS-SECTION
AT 8'

$F_x = 4,000 \text{ lb}$ $M_x = 4,800 \text{ in·lb}$

$F_y = -200 \text{ lb}$ $M_z = -19,200 \text{ in·lb}$

NOTE! THIS IS THE OUTWARD
FACE AT 8 ft

Point ①

$$\sigma_{AXIAL} = \frac{4000 \text{ lb}}{\pi(2\text{in})^2/4} = 1,273 \text{ psi}$$

$$\sigma_{BEND} = -\frac{(-19200 \text{ in·lb})(1\text{in})}{\pi(2)^4/64} = 24,446 \text{ psi}$$

$$\left. \begin{array}{c} \\ \\ \end{array} \right\} \sigma_{TOTAL} = 25,719 \text{ psi}$$

$$\tau_{TORQUE} = \frac{(4,800 \text{ in·lb})(1\text{in})}{\pi(2)^4/32} = 3,056 \text{ psi}$$

$$\tau_{BEND} = 0$$

$$A = \frac{1}{2}\pi r^2$$

$$\frac{4r}{3\pi} = \bar{y}$$

$$Q = A\bar{y} = \left(\frac{1}{2}\pi r^2\right)\left(\frac{4r}{3\pi}\right)$$

$$Q = \frac{2}{3}r^3$$

Point ②

$$\sigma_{AXIAL} = \frac{4000 \text{ lb}}{\pi(2\text{in})^2/4} = 1,273 \text{ psi}$$

$$\sigma_{BEND} = 0$$

$$\tau_{TORQUE} = \frac{(4,800 \text{ in·lb})(1\text{in})}{\pi(2)^4/32} = 3,056 \text{ psi}$$

$$\tau_{BEND} = \frac{(200)\left(\frac{2}{3}(1)^3\right)}{(\pi(2)^4/64)(2)} = 85 \text{ psi}$$

$$\left. \begin{array}{c} \\ \\ \end{array} \right\}$$ AT ② THESE ARE IN
THE SAME Direction

$$\tau_{TOTAL} = 3,141 \text{ psi}$$

Test Yourself 8.5

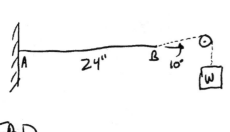

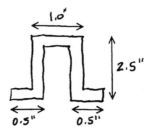

$\Sigma F_x = 0 \quad 1000\cos10° - F_x = 0 \quad \therefore \quad F_x = 984.8 \text{ lb}$

$\Sigma F_y = 0 \quad 1000\sin10 - F_y = 0 \quad \therefore \quad F_y = 173.7 \text{ lb}$

$\Sigma M_A = 0 \quad (1000\sin10)24 - M_z = 0 \quad \therefore \quad M_z = 4,169 \text{ in·lb}$

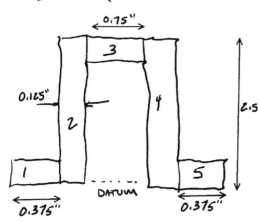

	(in) $\bar{y}_i$	(in²) $\bar{A}_i$	(in³) $\bar{y}_i \bar{A}_i$
1	0.0625	0.0468	0.00293
2	1.25	0.3125	0.391
3	2.4375	0.09375	0.229
4	1.25	0.3125	0.391
5	0.0625	0.0468	0.00293
Σ		0.8125 in²	1.017

$$\boxed{\bar{y} = \frac{1.017}{0.8125} = 1.25 \text{ in}}$$

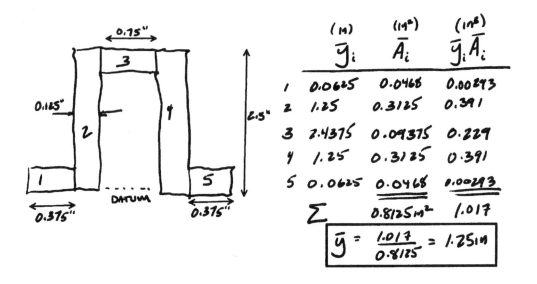

	(in) $\bar{y}_i$	(in²) $\bar{A}_i$	(in³) $\bar{y}_i \bar{A}_i$
1	0.0625	0.0468	0.00293
2	1.25	0.3125	0.391
3	2.4375	0.09375	0.229
4	1.25	0.3125	0.391
5	0.0625	0.0468	0.00293
Σ		0.8125 in²	1.017

$$\bar{y} = \frac{1.017}{0.8125} = 1.25 \text{ in}$$

Compute I_{zz}

$$I_{zz} = 2\left[\frac{1}{12}(0.375)(0.125)^3 + (0.0468)\left(1.25 - \frac{0.125}{2}\right)^2\right]$$

$$+ 2\left[\frac{1}{12}(0.125)(2.5)^3\right]$$

$$+ \frac{1}{12}(0.75)(0.125)^3 + (0.09375)(1.25 - 2.4375)^2$$

$$= 0.1321 + 0.3255 + 0.1323$$

$$\therefore \boxed{I_{zz} = 0.591 \text{ in}^4}$$

Axial Stress

$F_x = 984.8 \, lb$
$A = 0.8125 \, in^2$
$\left.\right\}$ $\sigma_{Axial} = \dfrac{984.8 \, lb}{0.8125 \, in^2} = 1,212 \, psi$

Bending Normal Stress

$M = 4,169 \, in \cdot lb$
$C = 1.25 \, in^2$
$I = 0.59 \, in^4$

$\sigma_{Bend} = \pm \dfrac{(4169)(1.25)}{0.59}$

$\sigma_{Bend} = \pm 8,833 \, psi$

Max Tensile Stress
$1,212 + 8,833 = 10,045 \, psi \quad (T)$

Max Compressive Stress
$1,212 - 8,833 = -7,621 \, psi \quad (c)$

SHEAR STRESS DUE TO BENDING

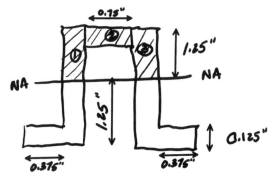

$\bar{y}_i$ (in)	A_i (in²)	$\bar{y}_i A_i$ (in³)
① $(1.25/2)$	$(1.25)(0.125)$	0.09766
② $(1.25 - 0.125/2)$	$(0.75)(0.125)$	0.1113
③ $(1.25/2)$	$(1.25)(0.125)$	0.09766
	$Q = \Sigma =$	0.3066

$$\tau_{Bend} = VQ/It = \frac{(F_y)(Q)}{I(2t)}$$

$$\tau_{Bend} = \frac{(173.7\,lb)(0.3066\,in^3)}{(0.59\,in^4)(2)(0.125\,in)} = 361 \ psi$$

Test Yourself 9.1

$\sigma_x = 4,000 \text{ psi}$

$\sigma_y = 0$

$\tau_{xy} = -500 \text{ psi}$

$\theta = 30° \text{ DOWNWARD}$

$\therefore \theta = -30°$

$2\theta = -60°$

$$\sigma_{x'} = \left(\frac{4000 + 0}{2}\right) + \left(\frac{4000 - 0}{2}\right)\cos(-60) + (-500)\sin(-60)$$

$$\sigma_{x'} = 2000 + 1000 + 433.0 = 3,433 \text{ psi}$$

$$\sigma_{y'} = \left(\frac{4000 + 0}{2}\right) - \left(\frac{4000 - 0}{2}\right)\cos(-60) - (-500)\sin(-60)$$

$$\sigma_{y'} = 2000 - 1000 - 433 = 567.0 \text{ psi}$$

$$\tau_{x'y'} = \left(\frac{0 - 4000}{2}\right)\sin(-60) + (-500)\cos(-60)$$

$$= 1732 - 250 = 1,482 \text{ psi}$$

Test Yourself 9.2

From the equation:

$$\sigma_1, \sigma_2 = \frac{\sigma_x + \sigma_y}{2} \pm \sqrt{\left(\frac{\sigma_x - \sigma_y}{2}\right)^2 + \tau_{xy}^2}$$

$$\therefore \boxed{\sigma_1 = \frac{45-18}{2} + \sqrt{\left(\frac{45+18}{2}\right)^2 + (9)^2} = 46.26 \, MPa}$$

$$\boxed{\sigma_2 = \frac{45-18}{2} - \sqrt{\left(\frac{45+18}{2}\right)^2 + (9)^2} = -19.26 \, MPa}$$

$$\tan 2\theta_p = \tau_{xy} / \left(\frac{\sigma_x - \sigma_y}{2}\right) = \tan 2\theta_p = \frac{9}{(45+18)/2} = 0.2857$$

$$\therefore \boxed{\theta_p = 7.97°} \quad (97.97°)$$

$$(\tau_{xy})_{MAX} = \pm \sqrt{\left(\frac{\sigma_x - \sigma_y}{2}\right)^2 + \tau_{xy}^2}$$

$$= \sqrt{\left(\frac{45+18}{2}\right)^2 + 9^2} = \boxed{32.76 \, MPa = (\tau_{xy})_{MAX}}$$

$$\tan 2\theta_s = -\frac{\sigma_x - \sigma_y}{2\tau_{xy}} = -\frac{45+18}{2(9)} = -3.5$$

$$\therefore \boxed{\theta_s = -37.03°} \quad (52.97°)$$

Test Yourself 9.3

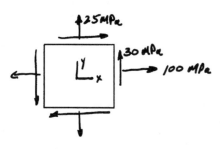

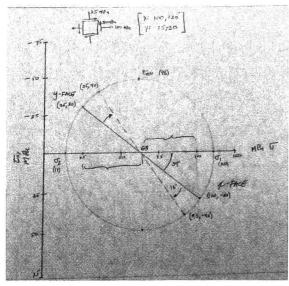

GRAPHICAL SOLUTION
(APPROX)

$\bar{\sigma} = 63$ MPa

$\sigma_1 = 110$ MPa $\sigma_2 = 14$ MPa

$\theta = 39°/2 = 18.5°$

$\tau_{MAX} = 48$ MPa

for $\theta = 8°$ $\sigma_x' = 90$ MPa $\sigma_y' = 25$ MPa
$\tau_{xy}' = 40$ MPa

USING Mohr's Circle As An AIDE

$\boxed{\bar{\sigma} = (100 + 25)/2 = 62.5 \text{ MPa}}$

$r = \sqrt{37.5^2 + 30^2} = 48.0$ MPa

$\boxed{\tau_{MAX} = r = 48 \text{ MPa}}$

$\bar{\sigma} + r = 62.5 + 48 = \boxed{110.5 \text{ MPa} = \sigma_1}$

$\sigma_2 = \bar{\sigma} - r = 62.5 - 48 = \boxed{14.5 \text{ MPa} = \sigma_2}$

$\tan\phi = \frac{30}{37.5} = 38.7°$ $\therefore \boxed{\theta = 19.35°}$

$90° - 38.7° - 16° = 35.3°$

$\boxed{\begin{array}{l}\sigma_x' = 62.5 + r\sin 35.3 = 90.2 \text{ MPa} \\ \sigma_y' = 62.5 - r\sin 35.3 = 34.8 \text{ MPa} \\ \tau_{xy}' = r\cos 35.3 = 39.2 \text{ MPa}\end{array}}$

Test Yourself 9.4

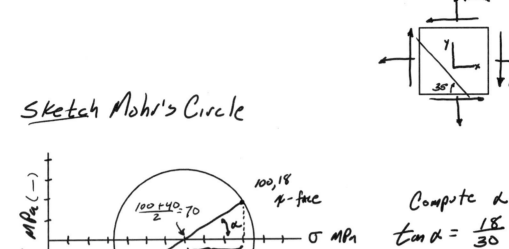

Sketch Mohr's Circle

Compute α

$\tan \alpha = \dfrac{18}{30}$

$\therefore \alpha = 30.96°$

NOTE: Radius of Mohr's Circle

$\sqrt{30^2 + 18^2} = 34.99 \text{ MPa}$

$\dfrac{100+40}{2} = 70$

$100 - 70 = 30$

100,18 x-face

40,18 y-face

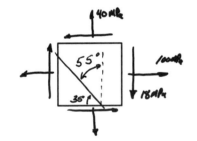

The x-face is Rotated 55° (110°)

$\therefore$ TOTAL: $30.96 + 110 = 140.96°$
on Mohr's Circle

AT 140.96°

$$\sigma_x' = 70 \text{ MPa} + 34.99 \text{ MPa} \cos(140.96)$$
$$\sigma_x' = 42.8 \text{ MPa}$$

$$\tau_{xy}' = -34.99 \sin(140.96)$$
$$\tau_{xy}' = -22.0 \text{ MPa}$$

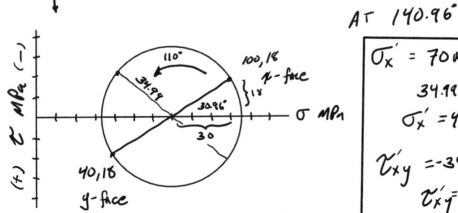

Test Yourself 9.5

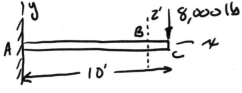

FBD - ABC

$\sum F_x = 0 \therefore N_A = 0$

$\sum F_y = 0 \quad V_A - 8000 = 0$

$\therefore V_A = 8000 \, lb$

$\sum M_A = 0 \quad -(8000\,lb)(10\,ft) + M_A = 0$

$\therefore M_A = 80,000 \, ft \cdot lb$

FBD AT CUT

$\sum F_y = 0 \quad 8,000 + V_B = 0 \quad \therefore \boxed{V_B = -8000\,lb}$

$\sum M_B = 0 \quad 80,000\,ft\,lb - (8000\,lb)(8\,ft) + M_B = 0$

$\therefore \boxed{M_B = -16,000 \, ft\,lb}$

AT B

$I = \frac{1}{12}(1.5)(4)^3 = 8.0 \, in^4$

$Q = (1.5)(1) \cdot (1.5) = 2.25 \, in^3$
Area · Dist.

$$\sigma_x = -\frac{(-16000\,ft\cdot lb)(12\,in/ft)(1\,in)}{8.0\,in^4} = 24,000\,psi$$

$$\tau_{xy} = \frac{(-8000\,lb)(2.25\,in^3)}{(8\,in^4)(1.5\,in)} = -1,500\,psi$$

At the cut:

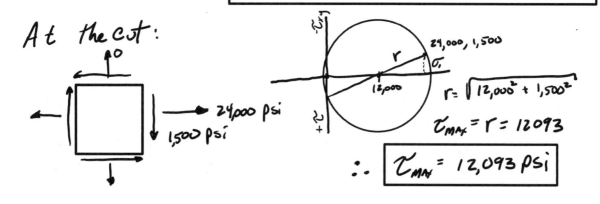

$r = \sqrt{12,000^2 + 1,500^2}$

$\tau_{max} = r = 12093$

$\therefore \boxed{\tau_{MAX} = 12,093 \, psi}$

Test Yourself 9.6

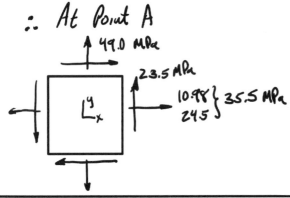

1,000N

$d_o = 30\,mm$

$t = 1\,mm$

3.5 MPa

30 N·M

1,000N

· A

① $\sigma_x = \dfrac{1000N}{\dfrac{\pi}{4}\left(0.030^2 - 0.028^2\right)m^2}$

$\boxed{\sigma_x = 10.98\ MPa}$ ①

② $\sigma_x = \dfrac{Pr}{2t} = \dfrac{(3.5 \times 10^6 Pa)(0.014m)}{2(0.001)m}$

$\boxed{\sigma_x = 24.5\ MPa}$ ②

③ $\sigma_y = \dfrac{Pr}{t} = \dfrac{(3.5 \times 10^6 Pa)(0.014m)}{(0.001\,m)}$

$\boxed{\sigma_y = 49.0\ MPa}$ ③

④ $\tau_{xy} = \dfrac{Tr}{J} = \dfrac{(30\ Nm)(0.015m)}{\dfrac{\pi}{2}(0.015^4 - 0.014^4)m} = 23.5\ MPa$

$\boxed{\tau_{xy} = 23.5\ MPa}$ ④

∴ At Point A

49.0 MPa

23.5 MPa

$\left.\begin{array}{c}10.98 \\ 24.5\end{array}\right\}$ 35.5 MPa

y
x

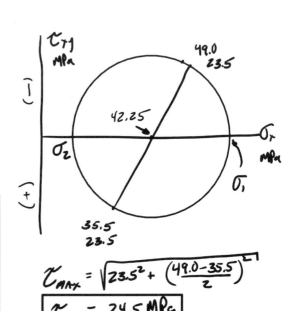

τ_{xy} MPa

49.0
23.5

42.25

σ_2

σ_r MPa

σ_1

35.5
23.5

$\tau_{max} = \sqrt{23.5^2 + \left(\dfrac{49.0 - 35.5}{2}\right)^2}$

$\boxed{\tau_{max} = 24.5\ MPa}$

Note: We NEGLECTED THE DIRECT STRESS IN THE Z-direction, WHICH WOULD Add A third PRINCIPAL STRESS. We'll see THAT IN THE NEXT EXAMPLE.

Test Yourself 9.7

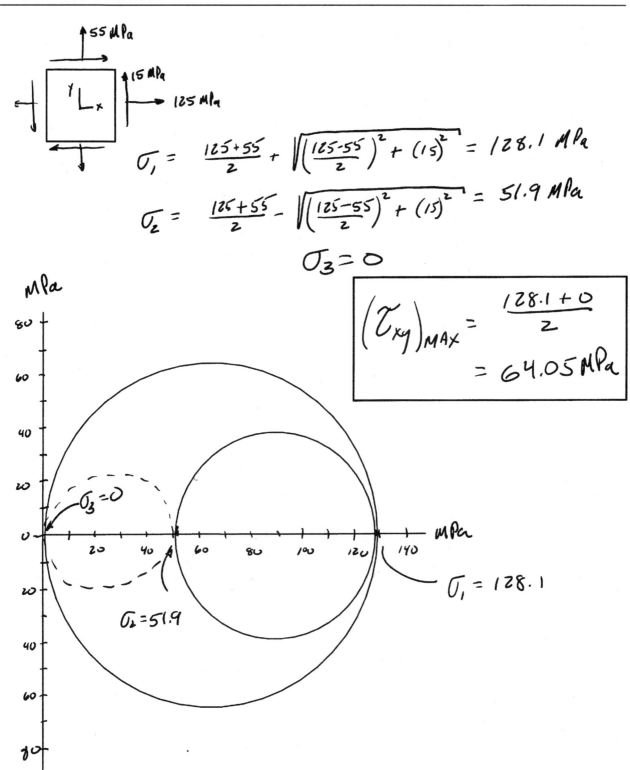

$$\sigma_1 = \frac{125+55}{2} + \sqrt{\left(\frac{125-55}{2}\right)^2 + (15)^2} = 128.1 \text{ MPa}$$

$$\sigma_2 = \frac{125+55}{2} - \sqrt{\left(\frac{125-55}{2}\right)^2 + (15)^2} = 51.9 \text{ MPa}$$

$$\sigma_3 = 0$$

$$(\tau_{xy})_{MAX} = \frac{128.1 + 0}{2} = 64.05 \text{ MPa}$$

Test Yourself 9.8

FOR A CYLINDRICAL PRESSURE VESSEL

$$\sigma_{HOOP} = \frac{pr}{t} = \frac{(5\times10^6 Pa)\big[(0.420m - 0.020m)\big]/2}{0.010\,m}$$

$$= 100\ MPa$$

$$\sigma_{LONG} = \frac{1}{2}\frac{pr}{t} = 50.0\ MPa$$

TRESCA

$$|\sigma_1| < \sigma_Y$$

$$100\,MPa \not< 95\,MPa \quad \underline{NO!}$$

$$\therefore FAILURE\ PREDICTED$$

VON MISES

$$\sqrt{\sigma_1{}^2 + \sigma_2{}^2 - \sigma_1\sigma_2} < \sigma_Y$$

$$\sqrt{100^2 + 50.0^2 - (100)(50.0)} = 86.6\ MPa$$

$$86.6\ MPa < 95\ MPa \quad \underline{OK}$$

$$FAILURE\ \underline{NOT}\ PREDICTED$$

Test Yourself 10.1

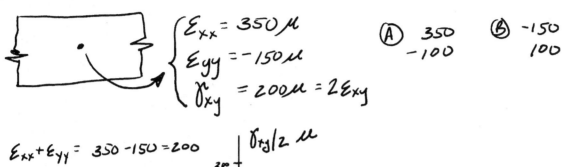

$$\varepsilon_{xx} = 350\,\mu$$
$$\varepsilon_{yy} = -150\,\mu$$
$$\gamma_{xy} = 200\,\mu = 2\varepsilon_{xy}$$

(A) 350
    −100

(B) −150
     100

$$\varepsilon_{xx} + \varepsilon_{yy} = 350 - 150 = 200$$
$$\varepsilon_{ANG} = \frac{200}{2} = 100\,\mu$$

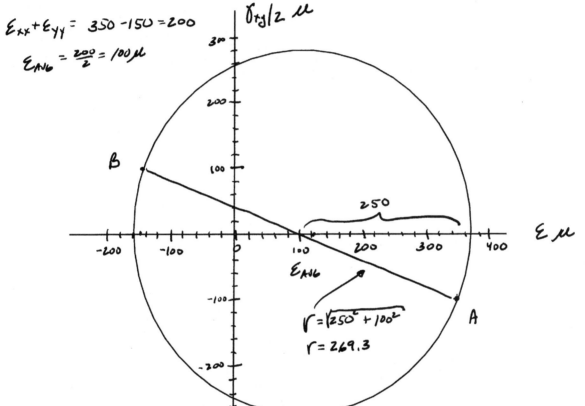

$$r = \sqrt{250^2 + 100^2}$$
$$r = 269.3$$

$$\varepsilon_1 = \varepsilon_{ANG} + r$$
$$= 100 + 269.3 = 369.3\,\mu$$
$$\varepsilon_2 = \varepsilon_{ANG} - r$$
$$= 100 - 269.3 = -169.3\,\mu$$
$$\left(\frac{\gamma_{xy}}{2}\right)_{MAX} = r = 269.3\,\mu$$
$$(\gamma_{xy})_{MAX} = 538.6\,\mu$$

Test Yourself 10.2

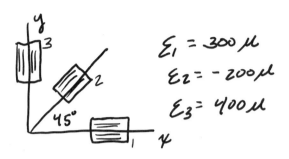

$$\varepsilon_1 = 300\mu$$
$$\varepsilon_2 = -200\mu$$
$$\varepsilon_3 = 400\mu$$

∴ $\varepsilon_x = 300\mu$

$\varepsilon_y = 400\mu$

$\varepsilon_{x'} = \varepsilon_2 = -200\mu$

$\theta_2 = 45°$

$$\varepsilon_2 = \varepsilon_x \cos^2\theta_2 + \varepsilon_y \sin^2\theta_2 + \gamma_{xy}\sin\theta_2\cos\theta_2$$

$$-200\mu = \frac{300\mu}{2} + \frac{400\mu}{2} + \frac{1}{2}\gamma_{xy}$$

∴ $\boxed{\gamma_{xy} = -1,100\mu}$

FIND PRINCIPAL STRAINS

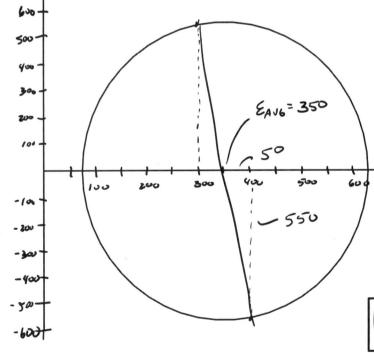

$\varepsilon_{AVG} = 350$

$\varepsilon_x = 300\mu$

$\varepsilon_y = 400\mu$

$\gamma_{xy}/2 = -550\mu$

$$\varepsilon_{AVG} = \frac{300+400}{2} = 350\mu$$

$$r = \sqrt{50^2 + 550^2}$$

$$r = 552.3\mu = (\varepsilon_{xy})_{max}$$

$$\boxed{(\gamma_{xy})_{MAX} = 1,104.5\mu}$$

Test Yourself 11.1

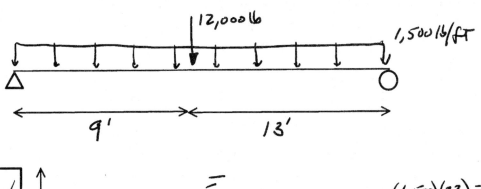

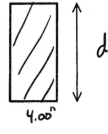

$$I = \frac{1}{12}(4)(d^3)$$

$$I = \frac{1}{3}d^3$$

$$\Sigma F_x = 0$$
$$\therefore \boxed{H_A = 0}$$

$$\Sigma M_A = 0$$
$$-(12,000)(9)\,lbft - (33,000)(11)\,lbft$$
$$+ 22\,R_b = 0$$
$$\therefore \boxed{R_b = 21,409.1\ lb}$$

$$\Sigma M_B = 0$$
$$0 = 22\,R_A - (12,000)(13)\,ft\,lb - (33,000)11\,ftlb$$
$$\therefore \boxed{R_A = 23,590.9\ lb}$$

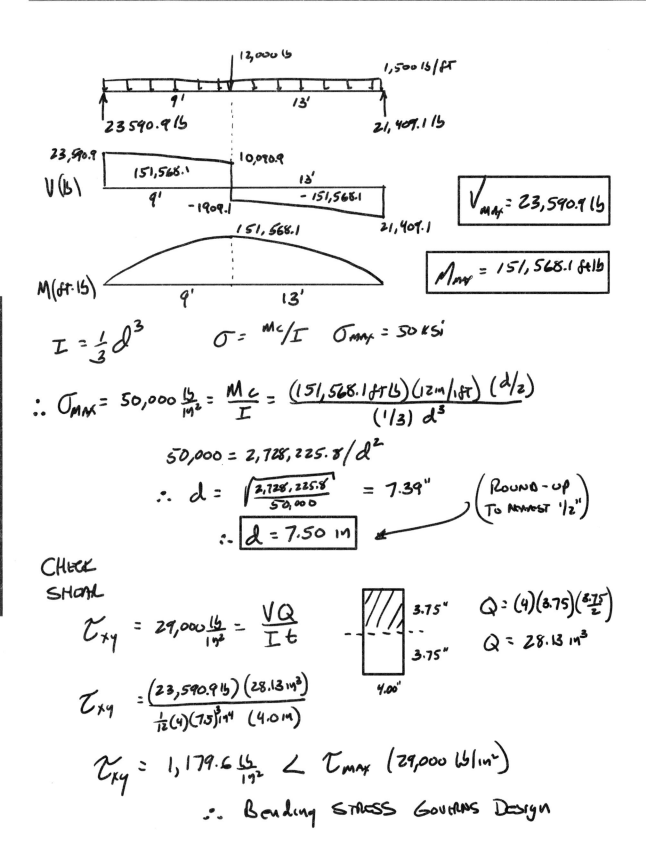

$$I = \frac{1}{3}d^3 \qquad \sigma = Mc/I \qquad \sigma_{MAX} = 50 \, ksi$$

$$\therefore \sigma_{MAX} = 50{,}000 \tfrac{lb}{in^2} = \frac{Mc}{I} = \frac{(151{,}568.1 \, ft\,lb)(12in/ft)\,(d/2)}{(1/3)\,d^3}$$

$$50{,}000 = 2{,}728{,}225.8 / d^2$$

$$\therefore d = \sqrt{\frac{2{,}728{,}225.8}{50{,}000}} = 7.39''$$

$$\left(\begin{array}{c} ROUND\text{-}UP \\ TO\ NEAREST\ 1/2'' \end{array} \right)$$

$$\therefore \boxed{d = 7.50 \, in}$$

CHECK
SHEAR

$$\tau_{xy} = 29{,}000 \tfrac{lb}{in^2} = \frac{VQ}{It}$$

$$Q = (4)(3.75)\left(\tfrac{3.75}{2}\right)$$

$$Q = 28.13 \, in^3$$

$$\tau_{xy} = \frac{(23{,}590.9 \, lb)(28.13 \, in^3)}{\frac{1}{12}(4)(7.5)^3 in^4 \,(4.0 in)}$$

$$\tau_{xy} = 1{,}179.6 \tfrac{lb}{in^2} < \tau_{MAX} \,(29{,}000 \, lb/in^2)$$

$$\therefore \text{ Bending STRESS GOVERNS Design}$$

Test Yourself 11.2

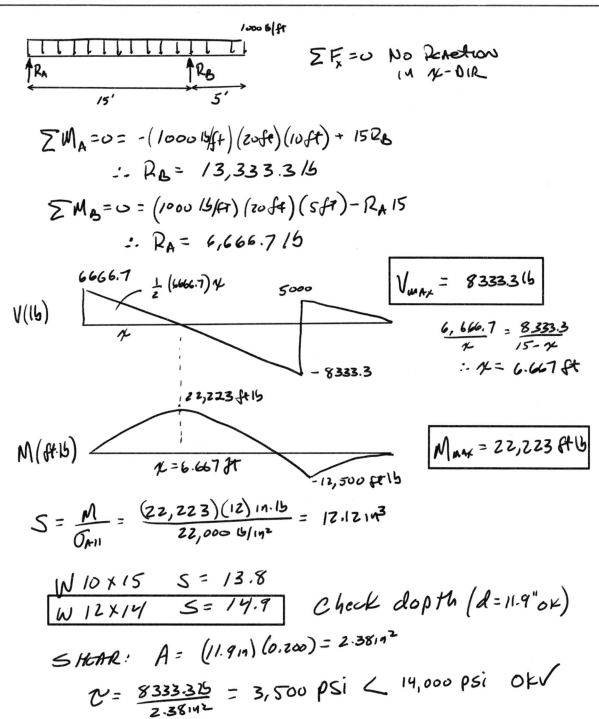

$$\sum F_x = 0 \quad \text{NO REACTION IN } x\text{-DIR}$$

$$\sum M_A = 0 = -(1000 \text{ lb/ft})(20 \text{ ft})(10 \text{ ft}) + 15 R_B$$

$$\therefore R_B = 13,333.3 \text{ lb}$$

$$\sum M_B = 0 = (1000 \text{ lb/ft})(20 \text{ ft})(5 \text{ ft}) - R_A 15$$

$$\therefore R_A = 6,666.7 \text{ lb}$$

$V(\text{lb})$

6666.7 $\frac{1}{2}(6666.7)x$ 5000

x

−8333.3

$$\boxed{V_{MAX} = 8333.3 \text{ lb}}$$

$$\frac{6,666.7}{x} = \frac{8333.3}{15-x}$$

$$\therefore x = 6.667 \text{ ft}$$

$M(\text{ft·lb})$

22,223 ft lb

$x = 6.667 \text{ ft}$

−12,500 ft lb

$$\boxed{M_{MAX} = 22,223 \text{ ft lb}}$$

$$S = \frac{M}{\sigma_{All}} = \frac{(22,223)(12) \text{ in·lb}}{22,000 \text{ lb/in}^2} = 12.12 \text{ in}^3$$

W 10×15	S = 13.8
W 12×14	S = 14.9

Check depth $(d = 11.9" \text{ ok})$

SHEAR: $A = (11.9 \text{ in})(0.200) = 2.38 \text{ in}^2$

$$\tau = \frac{8333.3 \text{ lb}}{2.38 \text{ in}^2} = 3,500 \text{ psi} < 14,000 \text{ psi} \quad \text{OK} \checkmark$$

Test Yourself 12.1

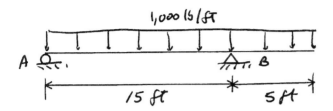

1,000 lb/ft

A ⚬ ⚬ ⚬ ⚬ ⚬ B

15 ft   5 ft

① REACTIONS

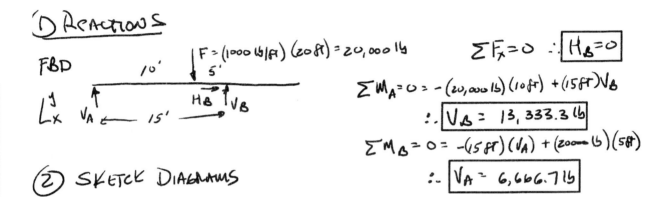

FBD

$F = (1000 \text{ lb/ft})(20 \text{ ft}) = 20,000 \text{ lb}$

$\sum F_x = 0 \quad \therefore \boxed{H_B = 0}$

$\sum M_A = 0 = -(20,000 \text{ lb})(10 \text{ ft}) + (15 \text{ ft}) V_B$

$\therefore \boxed{V_B = 13,333.3 \text{ lb}}$

$\sum M_B = 0 = -(15 \text{ ft})(V_A) + (2000 \text{ lb})(5 \text{ ft})$

$\therefore \boxed{V_A = 6,666.7 \text{ lb}}$

② SKETCH DIAGRAMS

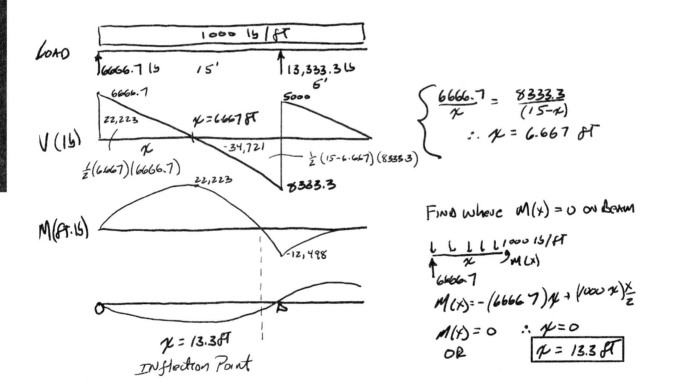

LOAD

1000 lb/ft

6666.7 lb   15'   13,333.3 lb

V (lb)

6666.7 $x = 6667 \text{ ft}$ 5000

22,223 x -34,721

$\frac{1}{2}(6667)(6666.7)$

$\frac{1}{2}(15 - 6.667)(83333)$

$\left\{ \begin{array}{l} \dfrac{6666.7}{x} = \dfrac{8333.3}{(15-x)} \\[2mm] \therefore x = 6.667 \text{ ft} \end{array} \right.$

M (ft·lb)

22,223

8333.3

-12,498

0   B

$x = 13.3 \text{ ft}$

INflection Point

FIND WHERE $M(x) = 0$ ON BEAM

1000 lb/ft

6666.7 x $M(x)$

$M(x) = -(6666.7)x + (1000\,x)\dfrac{x}{2}$

$M(x) = 0 \quad \therefore x = 0$

OR $\boxed{x = 13.3 \text{ ft}}$

Test Yourself 12.2

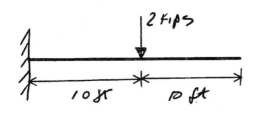

$\sum F_x = 0 \quad \boxed{H_A = 0}$

$\sum F_y = 0 \quad \boxed{V_A = 2 \text{ kips}} \quad \sum M_A = 0 = M_A - (2 \text{ kips})(10\text{ft})$

$\boxed{M_A = 20 \text{ kip·ft}}$

$\boxed{0 \leq x \leq 10\text{ft}}$

$\sum M = M + 20 - 2x = 0$

$\quad M(x) = 2x - 20$

$\dfrac{d^2 y_1}{dx^2} = \dfrac{M(x)}{EI} = \dfrac{1}{EI}(2x - 20)$

$\dfrac{dy_1}{dx} = \dfrac{1}{EI}\left(x^2 - 20x + C_1\right)$

$y_1 = \dfrac{1}{EI}\left(\dfrac{1}{3}x^3 - 10x^2 + C_1 x + C_2\right)$

$\boxed{10\text{ft} \leq x \leq 20\text{ft}}$

$\sum M = 20 - 2x + 2(x-10) + M = 0$

$\quad M = 0$

$\dfrac{d^2 y_2}{dx^2} = \dfrac{M}{EI} = 0$

$\dfrac{dy_2}{dx} = \dfrac{1}{EI}C_3$

$y_2 = \dfrac{1}{EI}(C_3 x + C_4)$

__BOUNDARY CONDITIONS:__

$x = 0 \quad y_1 = 0$

$x = 0 \quad \dfrac{dy_1}{dx} = 0$

__COMPATIBILITY CONDITIONS:__

$x = 10 \quad y_1 = y_2$

$x = 10 \quad \dfrac{dy_1}{dx} = \dfrac{dy_2}{dx}$

$$\therefore x=0 \quad y_1 = 0 = \frac{1}{EI}\left(\frac{1}{3}(0)^3 - 10(0)^2 + C_1(0) + C_2\right)$$

$$\therefore \boxed{C_2 = 0}$$

$$x=0 \quad \frac{dy_1}{dx} = 0 = \frac{1}{EI}\left((0)^2 - 20(0) + C_1\right)$$

$$\therefore \boxed{C_1 = 0}$$

$$x=10 \quad y_1' = y_2' \quad \frac{1}{EI}\left[(10)^2 - 20(10)\right] = \frac{1}{EI}\left[C_3\right]$$

$$100 - 200 = C_3 \implies \therefore \boxed{C_3 = -100.00}$$

$$x=10 \quad y_1 = y_2 \quad \frac{1}{EI}\left[\frac{1}{3}(10)^3 - 10(10)^2\right] = \frac{1}{EI}\left[-100.00(10) + C_4\right]$$

$$333.33 - 1000 = -1000 + C_4 \quad \therefore \boxed{C_4 = 333.33}$$

DISPLACEMENT AT THE TIP ($x = 20\,ft$)

$$y_2(20) = \frac{1}{EI}\left[-100(20) + 333.33\right]$$

$$\therefore \boxed{y_{TIP} = \frac{-1666.67}{EI}}$$

UNITS! We USED KIPS ; ft
SO THE DISPLACEMENT IS IN __ft__

$$IF \quad E = 29 \times 10^6 \, psi \left(\frac{1\,KIP}{1000\,lb}\right)\left(\frac{(12\,in)^2}{(1\,ft)^2}\right) = 4,176,000 \, KIPS/FT^2$$

$$I = 1401\,in^4 \left(\frac{(1\,ft)^4}{(12\,in)^4}\right) = 0.00675 \, ft^4$$

$$\therefore y_{TIP} = \frac{-1666.67}{(4176000)(0.00675)} = -0.059 \, ft$$

$$\boxed{y_{TIP} = -0.709 \, in}$$

Test Yourself 12.3

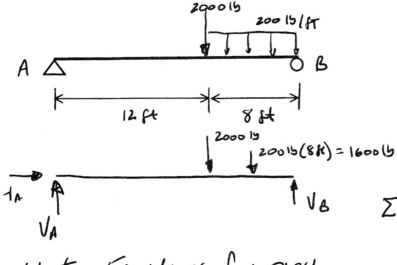

NOTE: we can write
the moment equation
By finding the reaction
on the left support.

Write Equations for each
Section using V_A

$0 \leq x \leq 12\,ft$

$\sum M_b = 0$

$-V_A(20\,ft) + 2000\,lb\,(8\,ft)$

$\quad + 1600\,lb\,(4\,ft) = 0$

$\therefore \boxed{V_A = 1,120\,lb}$

$\sum M_{cut} = 0 \quad M - 1120\,lb\,·\,x\,ft = 0$

$\therefore M = 1120\,x \quad ft·lb$

$\dfrac{d^2 y_1}{dx^2} = \dfrac{1}{EI}\left[1120x\right]$

$\dfrac{dy_1}{dx} = \dfrac{1}{EI}\left[560x^2 + C_1\right]$ ①

$y_1 = \dfrac{1}{EI}\left[\dfrac{560}{3}x^3 + C_1 x + C_2\right]$ ②

$12 \leq x \leq 20$

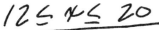

$2000\,lb$

$\left(200\frac{lb}{ft}\right)(x-12)\,ft$

$12'$

$(x-12)$

$\frac{(x-12)}{2}$

$1120\,lb$

$(x-12)$

M

V

x

$\sum M_{cut} = -(1120\,lb)(x\,ft) + (2000\,lb)(x-12)\,ft +$

$$+ \left[\left(200\frac{lb}{ft}\right)(x-12)\,ft\right]\frac{(x-12)}{2} + M = 0$$

$$\Rightarrow -1120x + 2000(x-12) + 200\frac{(x-12)^2}{2} + M = 0$$

$$\therefore M = 1120x - 2000(x-12) - 200\frac{(x-12)^2}{2}$$

$$\frac{d^2 y_2}{dx^2} = \frac{1}{EI}\left[1120x - 2000(x-12) - 200\frac{(x-12)^2}{2}\right]$$

$$\frac{dy_2}{dx} = \frac{1}{EI}\left[560x^2 - \frac{2000(x-12)^2}{2} - \frac{200(x-12)^3}{6} + C_3\right] \quad ③$$

$$y_2 = \frac{1}{EI}\left[\frac{560x^3}{3} - \frac{2000(x-12)^3}{6} - \frac{200(x-12)^4}{24} + C_3 x + C_4\right] \quad ④$$

BOUNDARY CONDITIONS:

$x=0 \quad y_1 = 0$

$x=20 \quad y_2 = 0$

COMPATIBILITY

$x=12 \quad y_1 = y_2$

$x=12 \quad dy_1/dx = dy_2/dx$

$$x=0 \quad y_1=0 \qquad 0=\frac{1}{EI}\left[\frac{560}{3}(0)^3+C_1(0)+C_2\right] \therefore \boxed{C_2=0}$$

$$x=20 \quad y_2=0$$

$$y_2=\frac{1}{EI}\left[\frac{560(20)^3}{3}-\frac{2000(8)^3}{6}-\frac{200(8)^4}{24}+C_3(20)+C_4\right]=0$$

$$\boxed{1.28853\times10^6+20C_3+C_4=0}\ (a)$$

COMPATIBILITY CONDITIONS: $\quad \underline{y_1=y_2 \quad x=12\ ft}$

$$\frac{1}{EI}\left[\frac{560}{3}(12)^3+C_1(12)+\cancel{C_2}^{0}\right]=$$

$$\frac{1}{EI}\left[\frac{560(12)^3}{3}-\frac{2000\cancel{(x-12)^3}^{0}}{6}-\frac{200\cancel{(x-12)^4}^{0}}{24}+C_3(12)+C_4\right]$$

$$\boxed{12C_1=12C_3+C_4}\ (b)$$

$$\underline{x=12 \quad y_1'=y_2'}$$

$$\frac{1}{EI}\left[560(12)^2+C_1\right]=$$

$$\frac{1}{EI}\left[560(12)^2-\frac{2000\cancel{(x-12)^2}^{0}}{2}-\frac{200\cancel{(x-12)^3}^{0}}{6}+C_3\right]$$

$$\therefore \boxed{C_1=C_3}\ (c)$$

Collect Equations ⓐ ⓑ ⓒ

ⓐ $\boxed{1.28853 \times 10^6 + 20C_3 + C_4 = 0}$

ⓑ $\boxed{12C_1 = 12C_3 + C_4}$

ⓒ $\boxed{C_1 = C_3}$

3 EQNS
3 UNKS
C_1, C_3, C_4
$(C_2 = 0)$

Combine ⓑ & ⓒ $12\cancel{C_3} = 12\cancel{C_3} + C_4$ ∴ $\boxed{C_4 = 0}$

From ⓐ $1.28853 \times 10^6 + 20C_3 = 0$ ∴ $\boxed{C_3 = -64426.7}$

From ⓒ   ∴ $\boxed{C_1 = -64426.7}$

$\underline{0 \leq x \leq 12}$

$y_1 = \frac{1}{EI}\left[\frac{560}{3}x^3 + C_1 x + C_2\right] = \frac{1}{EI}\left[\frac{560}{3}x^3 - 64426.7x\right]$

With $x = 10$ ft $y_1 = \frac{1}{EI}\left[\frac{560}{3}(10)^3 - 64,426.7(10)\right]$

$\boxed{y(10') = -\dfrac{457600}{EI}}$ (ft)

NOTE: $E = 29 \times 10^6$ psi
$= 4.176 \times 10^9 \frac{lb}{ft^2}$

$I = 140 \, in^4$
$= 0.00675 \, ft^4$

$\boxed{y_{10'} = \dfrac{-457600}{(4.176 \times 10^9)(0.00675)} = -0.0162 \, ft \\ (-0.195 \, in)}$

Test Yourself 12.4

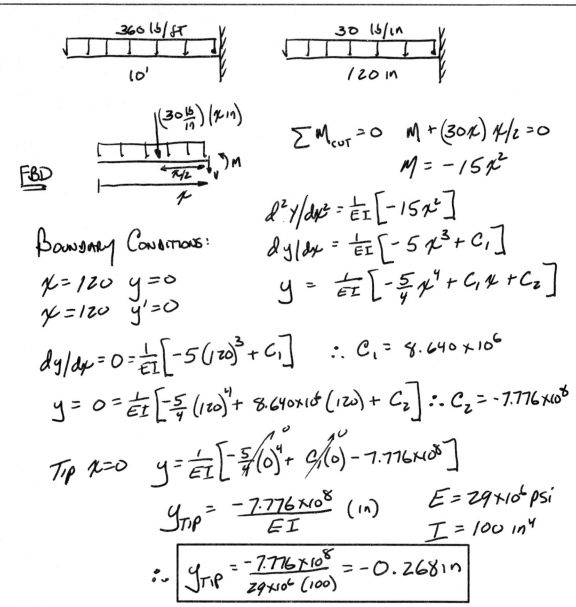

$$\sum M_{CUT} = 0 \quad M + (30x)\, x/2 = 0$$

$$M = -15x^2$$

$$d^2y/dx^2 = \frac{1}{EI}\left[-15x^2\right]$$

$$dy/dx = \frac{1}{EI}\left[-5x^3 + C_1\right]$$

$$y = \frac{1}{EI}\left[-\frac{5}{4}x^4 + C_1 x + C_2\right]$$

BOUNDARY CONDITIONS:

$x = 120 \quad y = 0$

$x = 120 \quad y' = 0$

$$dy/dx = 0 = \frac{1}{EI}\left[-5(120)^3 + C_1\right] \quad \therefore C_1 = 8.640 \times 10^6$$

$$y = 0 = \frac{1}{EI}\left[-\frac{5}{4}(120)^4 + 8.640 \times 10^6 (120) + C_2\right] \quad \therefore C_2 = -7.776 \times 10^8$$

TIP $x = 0$ $\quad y = \frac{1}{EI}\left[-\frac{5}{4}(0)^4 + C_1(0) - 7.776 \times 10^8\right]$

$$y_{TIP} = \frac{-7.776 \times 10^8}{EI} \text{ (in)} \qquad E = 29 \times 10^6 \text{ psi}$$

$$I = 100 \text{ in}^4$$

$$\therefore \boxed{y_{TIP} = \frac{-7.776 \times 10^8}{29 \times 10^6 (100)} = -0.268 \text{ in}}$$

Test Yourself 13.1

$$L = 8ft = 96in \qquad I_{MIN} = \frac{1}{12}(6in)(4in)^3 = 32in^4$$

$$E = 1.5 \times 10^6 \, psi$$

CASE I   Pinned-Pinned $k=1$

$$P_{CR} = \frac{\pi^2 EI}{(kL)^2} = \frac{\pi^2\left(1.5\times10^6 \frac{lb}{in^2}\right)(32\,in^4)}{(96in)^2}$$

$$\boxed{P_{CR} = 51,400 \, lbs}$$

CASE II   Fixed-Fixed $k=0.5$

$$P_{CR} = \frac{\pi^2 EI}{(kL)^2} = \frac{\pi^2\left(1.5\times10^6 \, lb/in^2\right)(32\,in^4)}{[(0.5)(96)]^2}$$

$$\boxed{P_{CR} = 206,000 \, lbs}$$

AS WE CAN SEE, THE BOUNDARY CONDITIONS MAKE A HUGE DIFFERENCE ON THE CRITICAL LOAD CAPACITY of The Column!

Test Yourself 13.2

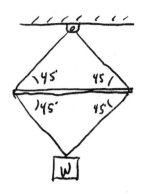

FBD

$$\sum F_y = 0 \quad 2(T\cos 45) - W = 0$$

$$1.414\,T = W$$

$$T = 0.707\,W$$

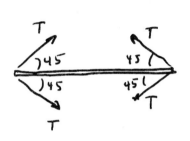

AT EACH END, WE HAVE:

$$2T\cos 45 = 2(0.707W)\cos 45 = W$$

BUCKLING: $\dfrac{\pi^2 E I}{(kL)^2}$ $k=1$, $E = 200\,GPa$ $I = \dfrac{\pi D^4}{64}$

$$P_{CR} = \pi^2 (200\times 10^9\,N/m^2)\left(\tfrac{\pi D^4}{64}\,m^4\right)/(1.5\,m)^2 \quad —① $$

YIELDING: $\sigma_y = -325\,MPa$ $A = \pi D^2/4$

$$P_{CR} = 325\times 10^6 \tfrac{N}{m^2} \cdot \tfrac{\pi D^2}{4}\,m^2 \quad —②$$

SET ① = ② SOLVE FOR D

$$(\pi^3)(200\times 10^9)D^4 / (64)(1.5)^2 = (\pi)(325\times 10^6)D^2/4$$

$$4.306\times 10^{10}\,D^4 = 2.553\times 10^8\,D^2 \Rightarrow D^2 = 5.928\times 10^{-3}\,m^2$$

$$\therefore D = 0.077\,M = 77\,mm \quad \left(\begin{array}{l}\text{NOTE: IGNORE } D=0 \\ \text{AND } D = -0.077\,m\end{array}\right)$$

$$W_{MAX} = (\sigma_y)A = (325\times 10^6 \tfrac{N}{m^2})(\pi(0.077)^2/4) = 1,513\,kN$$

Mechanics of Materials

Exam 1—Practice Set 1

Department of Mechanical Engineering
My Favorite University
Good Luck!

120 MINUTES
This class: Sec. 001 – Dr. Hanson

This exam covers Lesson 1–Lesson 22 (Level 1–Level 4)
Stress, Strain, Mechanical Properties, Axial Loads

RULES:

1. The solutions to this exam are video solutions which you can access at the QR code; however, you are not permitted to look at them until after your practice exam!
2. All your work must be done on the papers provided including this cover sheet.
3. Closed book and closed notes.
4. You may have a FE approved calculator, pens, pencils, erasers.
5. Use of a phone of any kind is considered cheating.
6. Treat this as your real exam and time yourself in a nice quiet place at a desk, as if you are taking this in your own classroom.

HONOR CODE—YOUR ME DEPARTMENT

I hereby certify that I will follow the Code of Student Conduct as defined by the University and the Department, that I will not cheat nor will I condone cheating.

Name _____ (Print legibly)

Name _____ (Signature)

PROBLEM 1 (20 POINTS)

Find the minimum diameter (d) of the pin at point B if:

a. $\tau_{allowable} = 45$ MPa
b. Bearing stress in the bell crank max is 95 MPa.
c. Bearing stress in support max is 155 MPa.

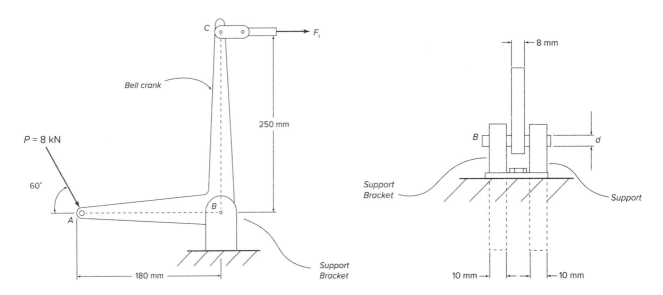

a. $d =$ _____

b. $d =$ _____

c. $d =$ _____

PROBLEM 2 (20 POINTS)

For the following find:

    a. The normal strain (ε) along line AC
    b. The normal strain (ε) along line BD
    c. The shear strain (γ) at corner A
    d. The shear strain (γ) at corner B
    e. The shear strain (γ) at corner C
    f. The shear strain (γ) at corner D

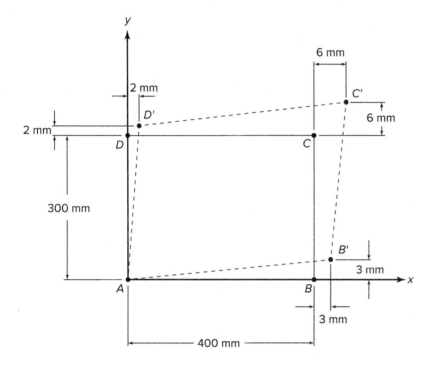

    a. $\varepsilon_{AC} =$ _____

    b. $\varepsilon_{BD} =$ _____

    c. $\gamma_A =$ _____

    d. $\gamma_B =$ _____

    e. $\gamma_C =$ _____

    f. $\gamma_D =$ _____

PROBLEM 3 (20 POINTS)

Find the following using the stress and strain diagram provided.

a. Modulus of elasticity

b. Maximum tensile force applied without permanent deformation for a 2.0 inch diameter rod. Use 0.2% offset rule and a factor of safety of 1.3.

c. How much will a 15' long, 2.53 inch diameter rod, subjected to 150 kips tensile load, elongate (i) while under load and (ii) after the load is removed?

d. If the sample is to absorb 555 in·# of energy without permanent deformation, what is the required volume?

e. If the sample is to absorb 555 in·# of energy without fracture, what is the required volume?

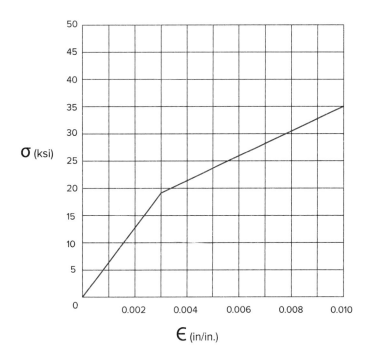

a. Elastic modulus = _____

b. Tensile force = _____

c. (i) While under load = _____ and
   (ii) After the load is removed = _____

d. Required volume (without permanent deformation) = _____

e. Required volume (without fracture) = _____

PROBLEM 4 (20 POINTS)

The steel pipe has an aluminum core bonded to it.

$E_{ST} = 200$ GPa $\qquad E_{AL} = 68.9$ GPa

   a. Find the average normal stress in aluminum.

   b. Find the average normal stress in steel.

   c. Find the total elongation of the assembly.

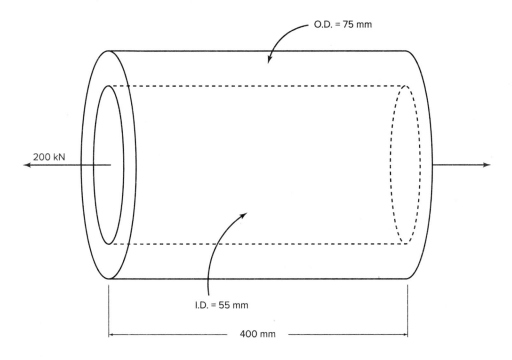

O.D. = 75 mm

200 kN

I.D. = 55 mm

400 mm

   a. Average normal stress in aluminum = _____

   b. Average normal stress in steel = _____

   c. Total elongation = _____

PROBLEM 5 (20 POINTS)

The steel bolt is tightened on the aluminum sleeve until the washers just make contact. The original temperature of 30°C is raised to 100°C. Find normal stress in the bolt and in the sleeve.

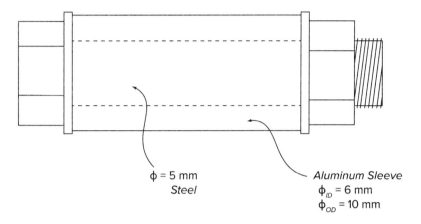

ϕ = 5 mm
Steel

Aluminum Sleeve
ϕ_{ID} = 6 mm
ϕ_{OD} = 10 mm

Material properties:

$E_{ST} = 200\ \text{GPa}$ $E_{AL} = 70\ \text{GPa}$

$\alpha_{ST} = 14 \times 10^{-6}/°C$ $\alpha_{AL} = 23 \times 10^{-6}/°C$

Normal stress in bolt = _____

Normal stress in sleeve = _____

Mechanics of Materials

Exam 1—Practice Set 2

Department of Mechanical Engineering
My Favorite University
Good Luck!

120 MINUTES

This class: Sec. 001 – Dr. Hanson

This exam covers Lesson 1–Lesson 22 (Level 1–Level 4)
Stress, Strain, Mechanical Properties, Axial Loads

RULES:

1. The solutions to this exam can be found in the back of the book; however, you are not permitted to look at them until after your practice exam.
2. All your work must be done on the papers provided including this cover sheet.
3. Closed book and closed notes.
4. You may have a FE approved calculator, pens, pencils, erasers.
5. Use of a phone of any kind is considered cheating.
6. Treat this as your real exam and time yourself in a nice quiet place at a desk, as if you are taking this in your own classroom.

HONOR CODE—YOUR ME DEPARTMENT

I hereby certify that I will follow the Code of Student Conduct as defined by the University and the Department, that I will not cheat nor will I condone cheating.

Name _____ (Print legibly)

Name _____ (Signature)

PROBLEM 1 (20 POINTS)

The wire *BD* has a yield stress of 250 MPa, $E = 200$ GPa, and Poisson's ratio of 0.3. Wire *BD* has a diameter of 8 mm.

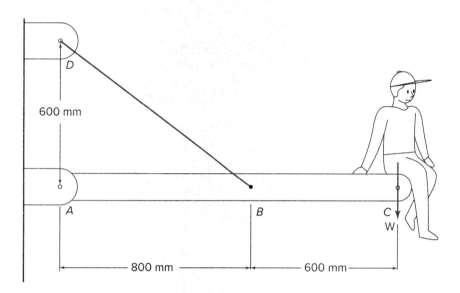

If the man has a mass of 100 kg, find the following:

a. The normal stress in wire *BD*
b. The factor of safety guarding against permanent deformation
c. The elongation of wire *BD* under load if angle *DAB* is 91.5°
d. The normal strain in wire *BD*
e. If the 5 mm pin at *A* is in double shear, find average shear stress on pin *A*

a. Normal stress in wire *BD* = _____

b. Factor of safety guarding against permanent deformation = _____

c. Elongation of wire *BD* = _____

d. Normal strain in wire *BD* = _____

e. Average shear stress on pin *A* = _____

PROBLEM 2 (20 POINTS)

The rectangular plate is deformed as shown.

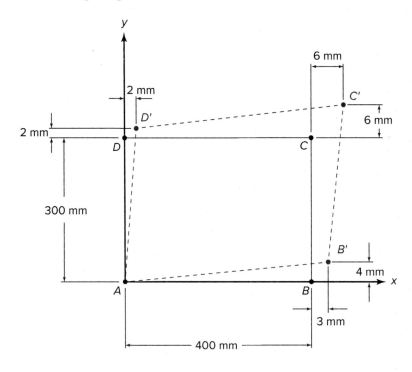

a. Find the average normal strain along diagonal *AC*.
b. Find the average normal strain along diagonal *BD*.
c. Find the average shear strain at corner *A*.
d. Find the average shear strain at corner *B*.

a. $\varepsilon_{AC} = $ _____

b. $\varepsilon_{BD} = $ _____

c. $\gamma_A = $ _____

d. $\gamma_B = $ _____

PROBLEM 3 (20 POINTS)

The stress-strain diagram for a material is shown here.

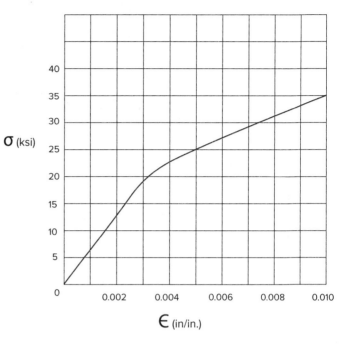

a. Find the Modulus of Elasticity.

b. Find the maximum allowable tensile force that can be applied if the specimen is a 2 in diameter rod of this material. Use the 0.2% offset rule and a factor of safety of 1.2.

c. If a 12 ft long rod having a diameter of 1.75 in was subjected to a tensile load of 80 kips, how much would it elongate while under load?

d. If the 12 ft rod with a 1.75 in diameter was subjected to a 80 kip tensile load and then unloaded, how much permanent deformation would it sustain?

e. If the material is to absorb impact energy of 500 in·# without permanent deformation, what would be the required volume of material? Use 0.2% offset rule for yield stress.

f. If the material is to absorb impact energy of 3 kip·in before fracturing, what would the minimum required volume be?

g. If a 2 in diameter rod of this material was subjected to a 50 kip tensile load, how much will the diameter decrease? (**Hint:** When subjected to an elastic shear stress of 18 ksi, the material sustains a corresponding shear strain of 0.009 rad.)

a. Elastic Modulus = _____

b. Tensile force = _____

c. While under load = _____

d. After the load is removed = _____

e. Required volume (without permanent deformation) = _____

f. Required volume (without fracture) = _____

g. Decreased diameter = _____

PROBLEM 4 (20 POINTS)

The assembly is constructed from the following materials:

- Bars *AB* and *CD* are copper ($E = 101$ GPa)
- Bar *EF* is stainless steel ($E = 193$ GPa)
- Consider the vertical bar in the middle of the assembly to be rigid.

a. Find the reaction at point *A*.
b. Find the average normal stress in *AB*.
c. Find the horizontal displacement of point *F* from the original.
d. Find the horizontal displacement of point *B* from the original.

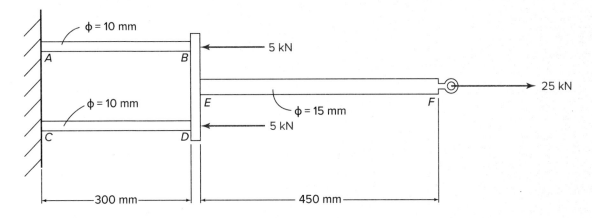

a. Reaction at point $A =$ _____

b. Average normal stress in $AB =$ _____

c. Horizontal displacement of point *F* from the original = _____

d. Horizontal displacement of point B from the original = _____

PROBLEM 5 (20 POINTS)

The shaft segments are between two walls. Section AB is hollow with a rigid cap at B, and section CD is solid. There is a 0.15 mm gap between the two sections; section AB is magnesium and CD is aluminum. If the temperature begins at 20°C, what is the highest temperature the system can experience before yielding failure?

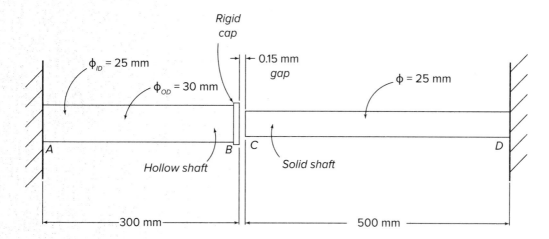

Material properties:

Section AB: Magnesium

$E = 44.7$ GPa

$\sigma_Y = 152$ MPa

$\nu = 0.30$

$\alpha = 26 \times 10^{-6}/°C$

Section CD: Aluminum

$E = 68.9$ GPa

$\sigma_Y = 255$ MPa

$\nu = 0.35$

$\alpha = 24 \times 10^{-6}/°C$

Highest temperature = _____

Mechanics of Materials

Exam 2—Practice Set 1

Department of Mechanical Engineering
My Favorite University
Good Luck!

120 MINUTES

This class: Sec. 001 – Dr. Hanson

This exam covers Lesson 23–Lesson 43 (Level 5–first part of Level 8)
Torsion, Bending, Transverse Shear, Thin-Walled Pressure Vessel

RULES:

1. The solutions to this exam are video solutions which you can access at the QR code; however, you are not permitted to look at them until after your practice exam!
2. All your work must be done on the papers provided including this cover sheet.
3. Closed book and closed notes.
4. You may have a FE approved calculator, pens, pencils, erasers.
5. Use of a phone of any kind is considered cheating.
6. Treat this as your real exam and time yourself in a nice quiet place at a desk, as if you are taking this in your own classroom.

HONOR CODE—YOUR ME DEPARTMENT

I hereby certify that I will follow the Code of Student Conduct as defined by the University and the Department, that I will not cheat nor will I condone cheating.

Name _____   (Print legibly)

Name _____   (Signature)

PROBLEM 1 (20 POINTS)

a. Find reactions at the walls, A and D.

b. Find shear stress (τ) in shafts AB and CD.

c. Find the angle of twist in gear B.

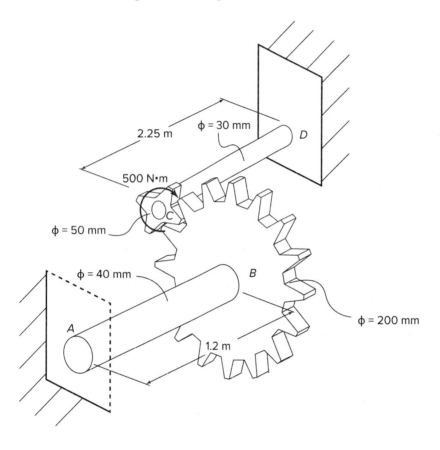

a. $T_A =$ _____ and $T_D =$ _____

b. $\tau_{AB} =$ _____ and $\tau_{CD} =$ _____

c. Angle of twist in gear $B =$ _____

PROBLEM 2 (20 POINTS)

Find the maximum tensile and compressive stresses in the beam. We will assume there is no bending out of plane (which is possible with a cross section like this). The vertical web has a thickness of 20 mm.

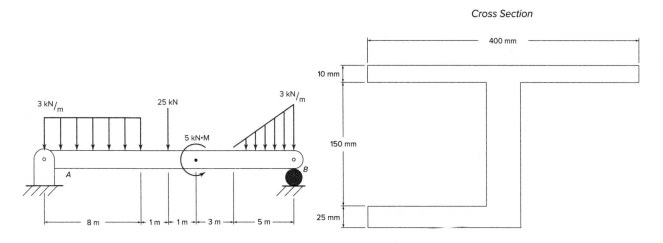

Cross Section

Maximum tensile stress = _____

Maximum compressive stress = _____

PROBLEM 3 (20 POINTS)

Find the maximum stress in the bonded beam shown in both the (a) steel and(b) wood.

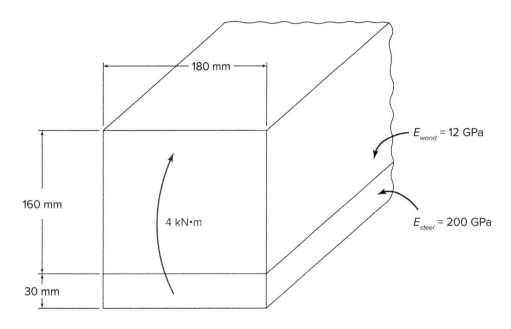

a. Maximum stress in steel = _____

b. Maximum stress in wood = _____

PROBLEM 4 (20 POINTS)

Find maximum P, given:

- Each nail can resist 2 kN.
- S (spacing of nails) = 120 mm
- Allowable shear stress of the beam (wood) = 3 MPa.

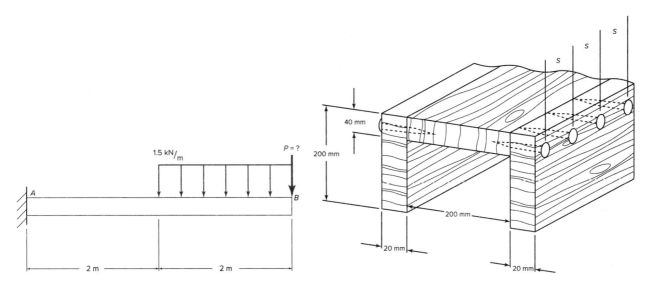

Max P = _____

PROBLEM 5 (20 POINTS)

An old mine site boiler is constructed as shown. The cylindrical portion of the body is made of two parts (i.e., two seams). The length of these horizontal seams are 12 m. If the tank must withstand 4 MPa, find the following:

a. The minimum thickness of the cylindrical body
b. The thickness of the hemispheric ends
c. The number of bolts required for the cylindrical seam
d. The number of bolts required for each end

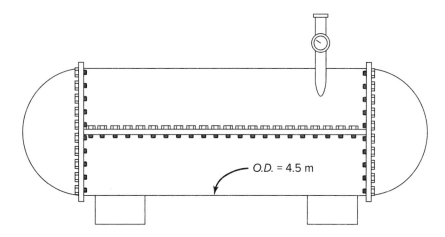

σ_{allow} for the tank = 150 MPa σ_{allow} for the bolts = 200 MPa

a. Minimum thickness in the cylinder = _____

b. Minimum thickness in the ends = _____

c. Number of bolt in the cylindrical seam = _____

d. Number of bolts for each end = _____

Mechanics of Materials

Exam 2—Practice Set 2

Department of Mechanical Engineering
My Favorite University
Good Luck!

120 MINUTES
This class: Sec. 001 – Dr. Hanson

This exam covers the end of Lesson 23–Lesson 43
(Level 5–first part of Level 8)

Torsion, Bending, Transverse Shear, Thin-Walled Pressure Vessel

RULES:

1. The solutions to this exam can be found in the back of the book; however, you are not permitted to look at them until after your practice exam.
2. All your work must be done on the papers provided including this cover sheet.
3. Closed book and closed notes.
4. You may have a FE approved calculator, pens, pencils, erasers.
5. Use of a phone of any kind is considered cheating.
6. Treat this as your real exam and time yourself in a nice quiet place at a desk, as if you are taking this in your own classroom.

PROBLEM 1 (20 POINTS)

A 2 in diameter solid steel shaft is used to transmit power from a motor to which it is attached. The steel has an allowable shear stress of 50 ksi, and the maximum angle of twist permitted in the design is 5°. The shear modulus of steel is 11×10^6 psi.

If the shaft rotates at 200 rpm and is 5 ft in length, what is the maximum hp motor that can be used for this application assuming that an FoS of at least 2 is desired for both the shear stress and the angle of twist?

Maximum hp = _____

PROBLEM 2 (20 POINTS)

A stepped shaft ACB is rigidly fixed between two supports. AC is solid steel and CB is composed of two materials. The inner core of CB is brass (material 1), and the outer shell of CB is aluminum (material 2). Assume $L_{AC} = 2$ m and $L_{CB} = 6$ m, the radius of AC is 25 mm, the inner diameter of CB is 25 mm, and the outer diameter of CB is 35 mm. The torque applied (T_o) is 30 kN·m.

What is the maximum torque in each segment of the shaft?

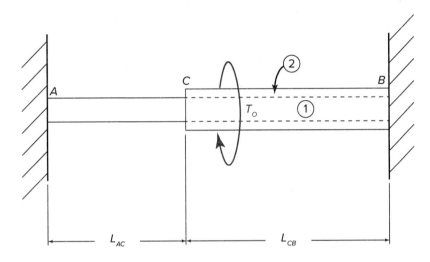

Material properties:

$E_{ST} = 200$ GPa	$E_{AL} = 70$ GPa	$E_{BRASS} = 101$ GPa
$G_{ST} = 75$ GPa	$G_{AL} = 27$ GPa	$G_{BRASS} = 37$ GPa

Maximum torque in $AC =$ _____

Maximum torque in $CB =$ _____

PROBLEM 3 (20 POINTS)

Use the beam loading below at a location of 8 ft from the left end of the beam to answer the following questions:

    a. What is the magnitude of the normal stress at point B?

    b. Is the normal stress tensile or compressive at point B?

    c. What is the shear stress at point B?

    d. What is the normal stress at point A?

    e. What is the shear stress at point A?

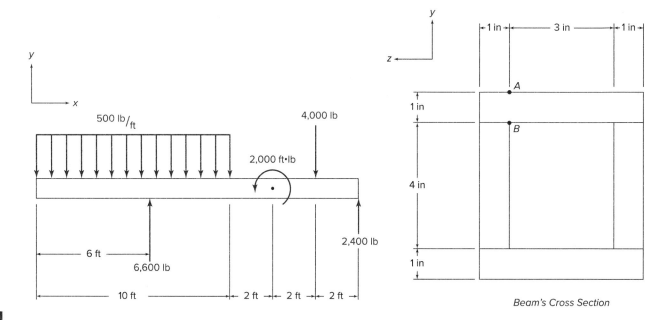

Beam's Cross Section

    a. Normal stress at point $B =$ _____

    b. The normal stress is in _____

    c. Shear stress at point $B =$_____

    d. Normal stress at point $A =$ _____

    e. Shear stress at point $A =$ _____

PROBLEM 4 (20 POINTS)

A thin-walled spherical pressure vessel has a 5.5 ft outer diameter and a ⅝ inch wall thickness with an internal pressure of 300 psi. If both are halves are bolted with a ½ inch diameter bolt that has a yield strength of 29 ksi, how many bolts would be required to bolt the two hemispherical halves together to withstand this pressure?

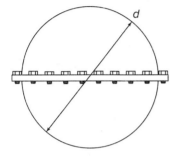

Number of bolts = _____

PROBLEM 5 (20 POINTS)

The following cross section is composed of pieces of brass and steel that are bonded together. The brass is on the top and the steel is on the bottom. If a moment of 4.5 kN·m is applied, what is the maximum bending stress in the:

   a. Brass

   b. Steel

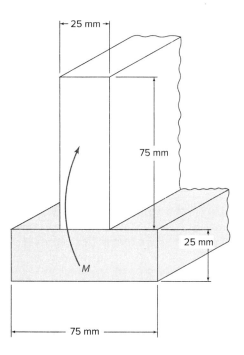

Material properties:

$E_{ST} = 200$ GPa $E_{AL} = 70$ GPa $E_{BRASS} = 101$ GPa

$G_{ST} = 75$ GPa $G_{AL} = 27$ GPa $G_{BRASS} = 37$ GPa

   a. Brass bending stress = _____

   b. Steel bending stress = _____

Mechanics of Materials

Exam 3—Practice Set 1

Department of Mechanical Engineering
My Favorite University
Good Luck!

120 MINUTES

This class: Sec. 001 – Dr. Hanson

This exam covers Lesson 44–Lesson 67 (second part of Level 8–Level 13)

Combined Loads, Stress and Strain Transformations, Rosettes, Beam Design, Slope and Deflection, and Buckling

RULES:
1. The solutions to this exam are video solutions which you can access at the QR code; however, you are not permitted to look at them until after your practice exam!
2. All your work must be done on the papers provided including this cover sheet.
3. Closed book and closed notes.
4. You may have a FE approved calculator, pens, pencils, erasers.
5. Use of a phone of any kind is considered cheating.
6. Treat this as your real exam and time yourself in a nice quiet place at a desk, as if you are taking this in your own classroom.

PROBLEM 1 (16.67 POINTS)

The solid pipe construction is 2 inches in diameter. Find the state of stress at point *A* and draw it on the stress element.

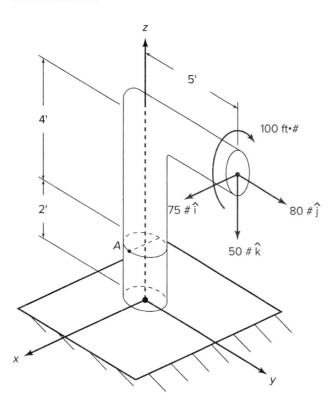

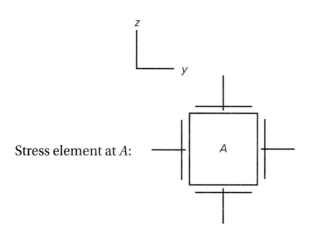

Stress element at *A*:

PROBLEM 2 (16.67 POINTS)

Find the following for the stress element below:

- a. Average normal stress
- b. Maximum shear stress
- c. Principal stresses
- d. Principal angles
- e. State of stress if rotated 25° clockwise

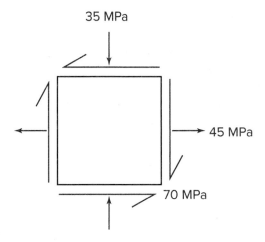

a. Average normal stress = _____

b. Maximum shear stress = _____

c. Principal stresses = _____

d. Principal angles = _____

e. State of stress if rotated 25° clockwise:

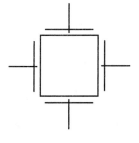

PROBLEM 3 (16.67 POINTS)

A strain rosette as shown below is placed on the backhoe's arm in order to determine its state of strain under loading. The three gauges read as follows: $A = 650 \times 10^{-6}$, $B = -320 \times 10^{-6}$, and $C = 500 \times 10^{-6}$. Determine the in-plane principal strains, and maximum in-plane shear strain.

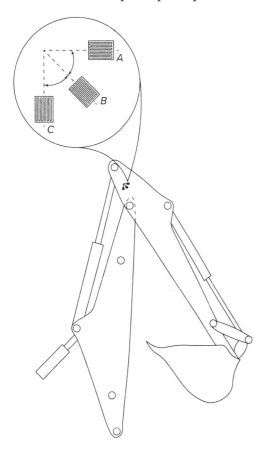

The in-plane principal strains = _____

Maximum in-plane shear strain = _____

PROBLEM 4 (16.67 POINTS)

The beam below is carrying a design load for a new construction project. The project manager needs help selecting the correct wide-flange beam for this application. If the allowable bending stress is $\sigma_{allow} = 22$ ksi and the allowable shear stress is $\tau_{allow} = 12$ ksi, select the lightest wide-flange beam that will carry this load. Use the table provided in the book.

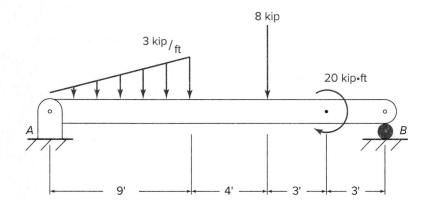

Wide-flange beam selection = _____

PROBLEM 5 (16.67 POINTS)

Determine the equations for slope and deflection for sections 1 and 2 of the loaded beam. E and I are constant.

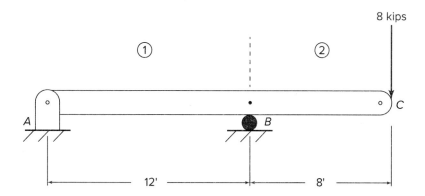

Slope equation (1) = _____

Slope equation (2) = _____

Deflection equation (1) = _____

Deflection equation (2) = _____

PROBLEM 6 (16.67 POINTS)

Determine the maximum distributed load (W) for the beam below to not buckle link BC. Link BC is comprised of steel and has a $E = 29,000$ ksi. ***Note:*** The top view of the link is shown and will cause the end connections to be pinned in one direction and fixed in another for buckling.

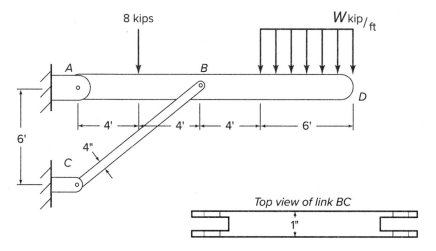

$W = $ _____

Exam 3—Practice Set 2

Department of Mechanical Engineering
My Favorite University
Good Luck!

120 MINUTES
This class: Sec. 001 – Dr. Hanson

This exam covers the end of Lesson 44–Lesson 67
(second part of Level 8–Level 13)

Combined Loads, Stress and Strain Transformations, Rosettes, Beam Design, Slope and Deflection, and Buckling

RULES:

1. The solutions to this exam can be found in the back of the book; however, you are not permitted to look at them until after your practice exam.
2. All your work must be done on the papers provided including this cover sheet.
3. Closed book and closed notes.
4. You may have a FE approved calculator, pens, pencils, erasers.
5. Use of a phone of any kind is considered cheating.
6. Treat this as your real exam and time yourself in a nice quiet place at a desk, as if you are taking this in your own classroom.

PROBLEM 1 (16.7 POINTS)

Which of the following stress elements correctly represents the principal stress state for the Mohr's circle shown?

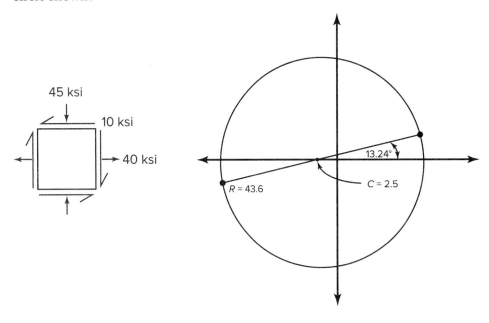

a)

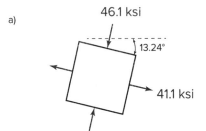

b)

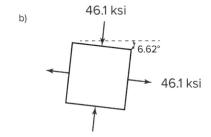

c)

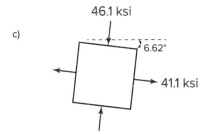

d)

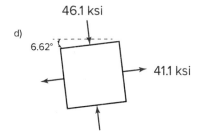

PROBLEM 2 (16.7 POINTS)

A pressurized cylindrical tank is fixed to the ground, and unfortunately exceptionally high winds are expected tonight. The wind load can be modeled as a point force halfway up the tank. Assume that the force produced by the wind is 150 kN, the inside diameter is 5 m, and the thickness of the tank is 2 cm. If the pressure in the tank is 0.2 MPa, what is the state of stress at point A?

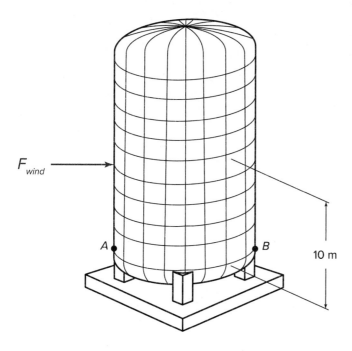

State of stress at point A: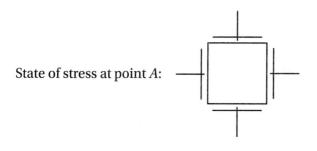

PROBLEM 3 (16.7 POINTS)

A beam has two point loads applied as shown. The beam is composed of brass, which has a Young's modulus of 14.5 Mpsi. The cross section is 2 inches by 4 inches.

For the following system, what is the deflection at B?

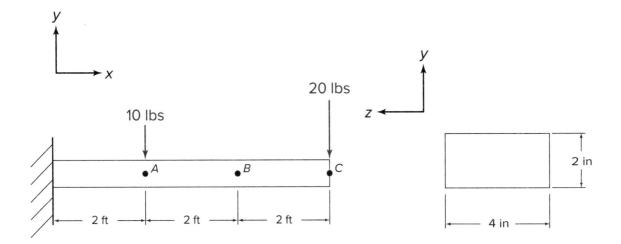

Deflection at $B =$ _____

PROBLEM 4 (16.7 POINTS)

The following beam has a cross section shown. If it is held in place by a guy wire, determine the maximum force (F) that can be applied without causing the system to buckle. Take $E = 29 \times 10^6$ psi and $\sigma_{yield} = 36.3 \times 10^3$ psi. Assume in the xy plane that the boundary conditions are pinned-pinned and in the zy plane it is fixed free.

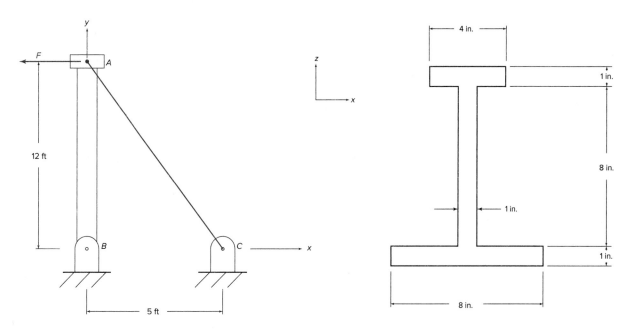

Maximum force $(F) =$ _____

PROBLEM 5 (16.7 POINTS)

A strain rosette is mounted to a wrench as shown. The following readings are obtained for the strain gauges, strain A $= -450 \times 10^{-6}$, strain B $= -600 \times 10^{-6}$, and strain C $= -300 \times 10^{-6}$. Find the:

a. In-plane principal strains
b. Maximum in-plane shear strain
c. The average normal strain

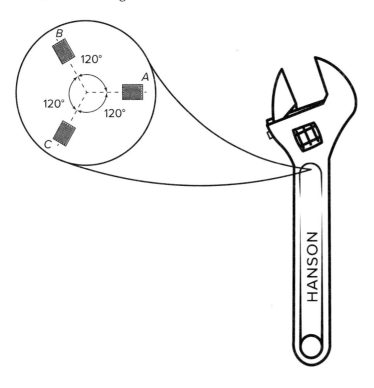

a. In-plane principal strains = _____

b. Maximum in-plane shear strain = _____

c. Average normal strain = _____

PROBLEM 6 (16.7 POINTS)

The pinned-beam is made from two sections, *AB* and *BC*. The beam supports two loads of 1.8 kips and 1.5 kips as shown. If the allowable bending stress is 24 ksi and the allowable shear stress is 14 ksi, select the lightest weight wide flange beam that will be safe to carry the load. Use the table provided in the book and a factor of safety of 1.5.

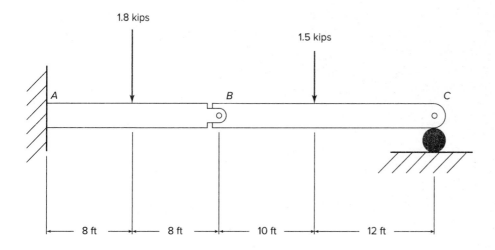

Wide-flange beam selection = _____

Exam
Solutions

EXAM 1, SET 2, PROBLEM 1

FBD

$W = (9.81 m/s^2)(100 kg)$
$= 981 N$

$\sum M_B = -A_y(0.8m) - (9.81)(0.6m) = 0$ ∴ $A_y = -735.75 N$

NOTE: θ for T_{BD}

$\tan^{-1}\left(\frac{6}{8}\right) = \theta$

$\theta = 36.87°$

$\sum F_y = 0$
$-735.75 - 981 + T_{BD} \sin\theta = 0$

∴ $T_{BD} = 2861.25 N$

$\sigma = F/A = T_{BD}/A_{BD} = \frac{2861.25 N}{(\pi/4)(0.008m)^2}$ ∴ $\sigma = 56.92 MPa$

$FoS = 250 MPa / 56.92 MPa = $ $FoS = 4.39$

NOTE: $L_{BD} = \sqrt{600^2 + 800^2} = 1000 MM$

$\Delta_{BD} = \frac{FL}{AE} = \frac{(2861.25N)(1.0m)}{(\pi/4)(0.008m)^2(200\times10^9 Pa)} = \Delta_{BD} = 0.285 mm$

$\varepsilon_{BD} = \frac{\Delta L_{BD}}{L_{BD}} = 0.285mm/1000mm = 2.85\times10^{-4} = \varepsilon_{BD}$

AT A: Double SHEAR $\phi = 5mm$

Reaction AT A $\sum F_x = 0 = 2861.25N \cos(36.87) + A_x$

∴ $A_x = -2288.997 N$

∴ Total Force → $\sqrt{(2288.997N)^2 + (735.75N)^2}$

$A = 2404.34 N$

$\tau = \frac{2404.34 N}{2 Area} = \frac{240434}{2(\pi/4)(0.005m)^2}$

$\tau = 61.23 MPa$

EXAM 1, SET 2, PROBLEM 2

(a) $\varepsilon_{AC} = \Delta L_{AC}/L_{AC}$
$L_{AC} = \sqrt{300^2 + 400^2} = 500mm$
$\Delta L_{AC} = (500 + \sqrt{6^2 + 6^2})mm - 500mm = 8.485mm$

$\varepsilon_{AC} = 8.485mm/500mm = 0.01697$

(b) $\varepsilon_{BD} = \Delta L_{BD}/L_{BD}$ (N.B. $L_{BD} = L_{AC} = 500mm$)
$L'_{BD} = \sqrt{(403-2)^2 + (4-302)^2} = 499.604 mm$

∴ $\Delta L_{BD} = 499.604 - 500 = -0.396 mm$
$\varepsilon_{BD} = \frac{0.396}{500} = -7.92\times10^{-4}$

(c)
$\tan^{-1}(2/300) = 0.0067$ radians
$\tan^{-1}(4/400) = 0.00999$ radians

$\gamma_A = \gamma_1 + \gamma_2 = 0.0167$ radians

(d)
$\gamma_1 = \tan^{-1}(3/403) = 0.00744$ radians
$\gamma_2 = \tan^{-1}(6/306) = 0.01961$ radians

∴ $\gamma = \gamma_1 + \gamma_2 = 0.02705$ radians

EXAM 1, SET 2, PROBLEM 3

ⓐ E Take Two Points $(15, 0.0024)$
$(0,0)$

$E = (15-0)/(0.0024-0) = 6250$ KSi

$$\boxed{E = 6.25 \text{ MPSi}}$$

ⓑ with $\varepsilon = 0.002$ (i.e., 0.2% offset) DRAW LINE

∴ Intercept $\simeq$ 27 KSi $Y \approx 27$ KSi

$F_oS = \dfrac{Y}{\sigma_{ACT}} = \dfrac{27 \text{ KSi}}{\sigma_{ACT}} = 1.2$

∴ $\sigma_{ACT} = F/A = 27 \text{ KSi}/1.2$ $F = \left(\dfrac{27 \text{ KSi}}{1.2}\right) \dfrac{\pi}{4}(2 \text{ in})^2$

$$\boxed{F = 70.7 \text{ KiP}}$$

ⓒ $\sigma = (80 \text{ KIPS})/(\pi/4)(1.75)^2 = 33.3$ KSi

$\varepsilon_{TOTAL} = 0.0083$ $\Delta L/L = \Delta L/144$

$$\boxed{\therefore \Delta L \simeq 1.2 \text{ in}}$$

ⓓ $\varepsilon_{TOTAL} - \varepsilon_{ELASTIC} = \varepsilon_{PLASTIC}$

$0.0083 - 0.0035 = 0.0048$

$\varepsilon = \Delta L/L \therefore \Delta L = \varepsilon L_o$

$\Delta L = 0.0048(144)$

$$\boxed{\Delta L = 0.69 \text{ in}}$$

ⓔ USE ONLY ELASTIC AREA UNDER CURVE

AREA $\simeq$ 13 BLOCKS; EACH BLOCK IS

$(5000 \text{ PSi})(0.001) = 5$ PSi

∴ Area $\simeq$ 5 PSi × 13 = 65 PSi

$U = 500 \text{ in·lb} = 65 \dfrac{lb}{in^2} \cdot \forall (in^3)$

$$\boxed{\forall = 7.7 \text{ in}^3}$$

ⓕ Area $\simeq$ 41 blocks   ∴ 5 PSi (41) = 205 PSi

$3000 \text{ in·lb} = 205 \dfrac{lb}{in^2} \cdot \forall$ $\boxed{\therefore \forall = 14.6 \text{ in}^3}$

ⓖ $G = \mathcal{T}/\gamma = \dfrac{18,000 \text{ PSi}}{0.009} = 2 \times 10^6$ PSi

$E = 2G(1+\nu) \Rightarrow 6.25 \times 10^6 = 2(2 \times 10^6)(1+\nu)$

∴ $\nu = 0.563$

$\varepsilon_d = -\nu \varepsilon_{length}$ $\Delta = FL/AE$ $\therefore \dfrac{\Delta}{L} = \dfrac{F}{AE} = \varepsilon$

$\varepsilon_{length} = \dfrac{(50,000 \text{ lb})}{\frac{\pi}{4}(2 \text{ in})^2 \cdot 6.25 \times 10^6 \text{ PSi}} = 2.55 \times 10^{-3}$ in/in

∴ $\varepsilon_d = -0.563(2.55 \times 10^{-3}) = -1.44 \times 10^{-3}$

$\dfrac{\Delta \text{ dia}}{\text{dia}} = -1.44 \times 10^{-3}$ ∴ $\Delta_{DIA} = -(2 \text{ in})(1.44 \times 10^{-3})$

$$\boxed{\Delta_{DIA} = -0.00288 \text{ in}}$$

EXAM 1, SET 2, PROBLEM 4

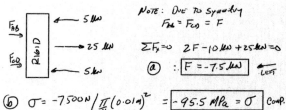

NOTE: DUE TO SYMMETRY
$F_{AB} = F_{CD} = F$

$\Sigma F_x = 0$ $2F - 10 \text{ kN} + 25 \text{ kN} = 0$

ⓐ ∴ $\boxed{F = -7.5 \text{ kN}}$ ← LEFT

ⓑ $\sigma = -7500 \text{ N}/\frac{\pi}{4}(0.01 \text{ m})^2 = \boxed{-95.5 \text{ MPa} = \sigma}$ COMP.

ⓒ $\Delta_F = \dfrac{(25,000)(0.45 \text{ m})}{\frac{\pi}{4}(0.015 \text{ m})^2 \cdot 193 \times 10^9 \text{ Pa}} - \dfrac{(7500 \text{ N})(0.30 \text{ m})}{\frac{\pi}{4}(0.010 \text{ m})^2 \cdot 101 \times 10^9 \text{ Pa}}$

$0.00033 \text{ M} - 0.00028 \text{ M}$

$$\boxed{\Delta_F = 0.00005 \text{ m} \ (0.05 \text{ mm})} \text{ right}$$

ⓓ From above

$\Delta_B = -\dfrac{(7500 \text{ N})(0.30 \text{ m})}{\frac{\pi}{4}(0.010 \text{ m})^2 \cdot 101 \times 10^9 \text{ Pa}} = -0.00028 \text{ M}$

$$\boxed{\Delta_B = -0.00028 \text{ m}} \text{ left}$$

EXAM 1, SET 2, PROBLEM 5

$$\left(F_{MAX}\right)_{Mg} = \sigma_y A = (152 \times 10^6 \, Pa) \frac{\pi}{4}\left(0.03^2 - 0.025^2\right)$$

$$\left(F_{MAX}\right)_{Mg} = 32.83 \, kN$$

$$\left(F_{MAX}\right)_{AL} = \sigma_y A = (255 \times 10^6 \, Pa) \frac{\pi}{4}(0.025)^2 = 125.2 \, kN$$

∴ MAGNESIUM Rod will YIELD FIRST AND CONTROL THE MAXIMUM TEMPERATURE.

$$\Delta = 0.15 \, mm = \left[\Delta_{TEMP} - \Delta_{FORCE}\right]_{AB} + \left[\Delta_{TEMP} - \Delta_{FORCE}\right]_{CD}$$

$$0.00015m = \left[\alpha L\Delta T - \frac{32,830 \, L}{A\,E}\right]_{AB} + \left[\alpha L\Delta T - \frac{32,830 \, L}{A\,E}\right]_{CD}$$

$$0.00015 = \left[26 \times 10^{-6}(0.30)\Delta T - \frac{32,830(0.3)}{\frac{\pi}{4}(0.03^2 - 0.025^2)(44.7 \times 10^9)}\right]_{AB}$$

$$+ \left[24 \times 10^{-6}(0.50)\Delta T - \frac{32,830(0.5)}{\frac{\pi}{4}(0.025)^2(68.9 \times 10^9)}\right]_{CD}$$

$$0.00015 = \left[7.8 \times 10^{-6}\Delta T - 1.02 \times 10^{-3}\right] + \left[1.2 \times 10^{-5}\Delta T - 4.853 \times 10^{-4}\right]$$

$$0.00015 = 1.98 \times 10^{-5}\Delta T - 1.505 \times 10^{-3}$$

$$\Delta T = 83.6 °C \qquad \text{BUT Ambient } 20°C$$

∴ HIGHEST TEMP $83.6 + 20 = 103.6$

$$\boxed{T_{MAX} = 103.6 \, °C}$$

EXAM 2, SET 2, PROBLEM 1

$$FoS = 2 \quad \therefore \quad \tau_{ALL} = \frac{50 \, ksi}{2} = 25 \, ksi \quad \theta_{ALL} = \frac{5°}{2} = 2.5°$$

$$\omega = \frac{200 \, rev}{min} \cdot \frac{1 \, min}{60 \, s} \cdot \frac{2\pi \, radian}{1 \, rev} = 20.94 \, radians/s$$

$$P = \omega T \qquad \tau_{ALL} = Tr/J \quad \therefore \quad T = \frac{\tau_{ALL} \cdot J}{r}$$

$$J = \frac{\pi d^4}{32} = \pi(2^4)/32 = 1.571 \, in^4$$

$$\therefore \quad T = (25,000 \, lb/in^2)(1.571 \, in^4)/1 \, in = 39,275 \, in \cdot lb$$

$$\therefore \quad P = \omega T = 20.94 \frac{radians}{s}(39,275 \, in \cdot lb) = 822419 \frac{in \, lb}{s}$$

NOTE: $1 \, hp = 550 \, ft \, lb/s$

$$P = 822,419 \frac{in \, lb}{s} \cdot \frac{1 \, ft}{12 \, in} = 68,535 \, ft \, lb/s$$

$$\therefore \quad P = 68,535 \, ft \cdot lb/s \cdot \frac{1 \, hp}{550 \, ft \, lb/s} = \boxed{124.6 \, hp}$$

$$\theta = TL/GJ \qquad \theta_{ALL} = 2.5° \cdot \frac{2\pi \, radian}{360°} = 0.04363 \, radians$$

$$T_{ALL} = \frac{\theta \, GJ}{L} = \frac{(0.04363)(11 \times 10^6)(1.571)}{(5)(12)} = 12,567 \, in \, lb$$

$$P = T\omega = (12567)(20.94) = 263,155 \, in \, lb/s$$

$$= 21929.6 \, ft \cdot lb/s$$

$$= 39.87 \, hp$$

$$\therefore \quad \boxed{P_{ALL} = 39.87 \, hp} \quad \swarrow \text{MAX ALLOWABLE LIMITED BY THE ANGLE OF TWIST}$$

EXAM 2, SET 2, PROBLEM 2

BRASS ①
ALUM ②
STEEL ③

$$J_1 = \frac{\pi}{32}\left[(2)(0.025m)\right]^4 = 6.136 \times 10^{-7} \, m^4$$

$$J_2 = \frac{\pi}{32}\left[(2)(0.035m)\right]^4 - \frac{\pi}{32}\left[(2)(0.025m)\right]^4 = 1.744 \times 10^{-6} \, m^4$$

$$J_3 = \frac{\pi}{32}\left[(2)(0.025m)\right]^4 = 6.136 \times 10^{-7} \, m^4$$

FBD:

$$\sum T = 0 \quad \therefore \; T_A + T_B = T_0 = 30,000 \; N \cdot m \quad —①$$

Also: $\phi_{AC} = \phi_{BC}$

$$\frac{T_A \, L_{AC}}{G_3 J_3} = \frac{T_B \, L_{CB}}{G_1 J_1 + G_2 J_2} \quad —②$$

Sub values into ②

$$\frac{T_A (2m)}{(75 \times 10^9 Pa)(6.136 \times 10^{-7} m^4)} = \frac{T_B (6m)}{(37 \times 10^9 Pa)(6.136 \times 10^{-7} m^4) + (27 \times 10^9 Pa)(1.744 \times 10^{-6} m^4)}$$

$$\frac{T_A (2m)}{(75 \times 10^9 Pa)(6.136 \times 10^{-7} m^4)} = \frac{T_B (6m)}{(37 \times 10^9 Pa)(6.136 \times 10^{-7} m^4) + (27 \times 10^9 Pa)(1.744 \times 10^{-6} m^4)}$$

$$4.346 \times 10^{-5} T_A = 8.597 \times 10^{-5} T_B \quad —②$$

$$\hookrightarrow 4.346 \, T_A = 8.597 \, T_B$$

$$T_A = \frac{8.597}{4.346} T_B = 1.978 \, T_B$$

From ① $\quad T_A + T_B = 30,000 \; N \cdot m$

$$1.978 \, T_B + T_B = 30,000$$

$$\therefore \boxed{\begin{array}{l} T_B = 10,074 \; NM \\ T_A = 1.978 \, T_B = 19,926 \; NM \end{array}}$$

EXAM 2, SET 2, PROBLEM 3

$$\sum M_{CUT} = 0 = -6600(2) + 4000(4) + M = 0$$

$$\therefore M = -2,800 \; ft \cdot lb$$

$$= -33,600 \; in \cdot lb$$

For X-SEC $\quad I = \frac{1}{12}(5)(6)^3 - \frac{1}{12}(3)(4)^3 = 74 \, in^4$

$$\sigma = -\frac{My}{I} = (33,600 \, in \cdot lb)(2 in) / 74 \, in^4 = 908.1 \, psi$$

a) $\boxed{\sigma = 908.1 \, psi}$ b) $\boxed{\text{Tension}}$

From FBD - compute V

$$\sum F_y = 0 \quad -V + 6600 \, lb - 4000 \, lb = 0 \quad \therefore V = 2,600 \, lb$$

for X-SEC: $\bar{y}$ AT 3in from base

Compute Q AT cut: $Q = \left[(5in)(1in)\right](3in - 0.5in) = 12.5 \, in^3$

$$\tau = \frac{VQ}{Ib} = \frac{(2,600 \, lb)(12.5 \, in^3)}{(74 \, in^4)(2 in)} = 219.6 \, psi$$

c) $\boxed{\tau = 219.6 \, psi}$

$$\sigma = -\frac{My}{I} = (33,600 \, in \cdot lb)(3 in) / 74 \, in^4 = 1,362 \, psi$$

d) $\boxed{\sigma = 1,362 \, psi}$ e) $\boxed{\text{zero}}$

EXAM 2, SET 2, PROBLEM 4

Inner Diameter: $66\text{in} - 2(5/8\text{in}) = 64.75\text{in}$

Total Force: $F = p \cdot A = \left(300 \frac{lb}{in^2}\right)\left(\frac{\pi}{4}(64.75in)^2\right)$

$$F = 987,849 \text{ lb}$$

Force per bolt

$\sigma_y = 29 \text{ KSi}$ $A_{bolt} = \frac{\pi (0.5)^2}{4} = 0.19635 \text{ in}^2$

$F_{bolt} = \left(29 \times 10^3 \frac{lb}{in^2}\right)\left(0.19635 in^2\right) = 5,694 lb$

\# bolts $= \frac{987,849 lb}{5,694 lb} = 173.5$ ∴ $\boxed{174 \text{ bolts}}$

EXAM 2, SET 2, PROBLEM 5

Convert beam to steel

$\eta = E_{steel}/E_{brass} = 200/101 = 1.98$

∴ New x-sec

Compute Centroid & I

$\bar{y} = \frac{(25)(75)\left(\frac{25}{2}\right) + (75)(12.63)(75/2 + 25)}{(25)(75) + (75)(12.63)} = 29.28 \text{ mm}$

$I = \frac{1}{12}(75)(25)^3 + (75)(25)(29.28 - 12.5)^2$
$+ \frac{1}{12}(12.63)(75)^3 + (12.63)(75)(75/2 + 25 - 29.28)^2 = 2.115 \times 10^6 \text{ mm}^4$
$\therefore I \approx 2.115 \times 10^6 \text{ m}^4$

$M = 4.5 \text{ kN·m} = 4,500 \text{ N·m}$

$\sigma_{ST} = \frac{(4500 N·m)(0.02928m)}{2.115 \times 10^{-6} m^4} = 62.3 \text{ MPa}$

$\sigma_{BRASS} = \frac{(-4500)(0.100 - 0.02928)}{2.115 \times 10^{-6}} = -150.5 \text{ MPa}$

Convert to original brass material:

$$\sigma_{BRASS} = (-150.5 \text{ MPa})/1.98 = 76.0 \text{ MPa}$$

EXAM 3, SET 2, PROBLEM 1

$$\sigma_1, \sigma_2 = \frac{\sigma_x + \sigma_y}{2} \pm \sqrt{\left(\frac{\sigma_x - \sigma_y}{2}\right)^2 + \tau_{xy}^2} \quad \text{ksi}$$

$$= \left(\frac{40-45}{2}\right) \pm \sqrt{\left(\frac{40--45}{2}\right)^2 + (-10)^2} \quad \text{ksi}$$

$$= -2.5 \pm 43.66$$

$$\sigma_1 = 41.16 \quad \checkmark$$
$$\sigma_2 = -46.16$$

$$\tan(2\theta_1) = \tau_{xy} / (\sigma_x - \sigma_y)/2 = -0.116 \text{ radians}$$
$$(\text{or } \theta_p = -6.62°)$$

From Mohr's Circle $2\theta = 13.24°$ ∴ $\theta = 6.62°$ (Negative)

Answer C)

EXAM 3, SET 2, PROBLEM 2

X-Direction

Hoop Stress $\sigma_H = \frac{pr}{t} = \frac{(0.2 \times 10 \text{ MPa})(5m/2)}{0.020 \text{ m}} = 25 \text{ MPa}$ Tensile

Y-Direction

Bending Stress $\sigma_B = \frac{Mc}{I}$

$$I = \frac{\pi}{4}\left[\left(\frac{5.04 \text{ m}}{2}\right)^4 - \left(\frac{5.00}{2}\right)^4\right] = 0.9936 \text{ m}^4$$

$$c = 5.04 \text{ m}/2 = 2.52 \text{ m}$$

$$\sigma_B = (150,000 \text{ N})(10 \text{ m})(2.52 \text{ m})/0.9936 \text{ m}^4$$
$$\sigma_B = 3.8 \text{ MPa} \quad \text{Tensile}$$

Longitudinal Stress (due to pressure)
$$\sigma_{long} = \frac{1}{2}\sigma_{Hoop} = 12.5 \text{ MPa} \quad (\text{Tensile})$$

S-perpose: $\sigma_x = 3.8 \text{ MPa} + 12.5 \text{ MPa} = 16.3 \text{ MPa}$

Stress State

EXAM 3, SET 2, PROBLEM 3

$$\sum F_y = 0$$
$$-10 - 20 + R_A = 0 \quad \therefore R_A = 30 \, lb \uparrow$$

$I = \frac{1}{12}(4)(2)^3 = 2.667 \, in^4$

$$\sum M_A = 0 \quad -(10)(2) - 20(6) + M_A = 0$$
$$\therefore M_A = 140 \, ft \cdot lb = 1680 \, in \cdot lb$$

Write the Moment Equation for $24 \le x \le 48 \, in$

$$\sum M_x = 0 \quad M(x) + (10 \, lb)(x-24)in - (30 \, lb)(x)in + 1680 \, in \cdot lb = 0$$

$$M(x) = -10(x-24) + 30x - 1680 \quad in \cdot lb$$
$$= -10x + 240 + 30x - 1680$$
$$M(x) = 20x - 1440$$

$$EI \frac{d^2y}{dx^2} = M(x) = 20x - 1440$$

$$EI \, dy/dx = 10x^2 - 1440x + C_1$$

$$EI \, y = \frac{10}{3}x^3 - 720x^2 + C_1 x + C_2$$

BC's:
$x = 0 \quad y = 0 \quad \therefore C_2 = 0$
$x = 0 \quad y' = 0 \quad \therefore C_1 = 0$

$$\therefore y_B = \frac{1}{EI}\left[\frac{10}{3}x^3 - 720x^2\right]_{x=48} = \frac{-1,290,240}{EI}$$

$$\therefore \boxed{y_B = -1,290,240 / (14.5 \times 10^6)(2.667) = -0.0334 \, in}$$

EXAM 3, SET 2, PROBLEM 4

NOTE: $xy \rightarrow$ Pinned-Pinned
$yy \rightarrow$ fixed-free

Compute I_{zz}

$$I_{zz} = \frac{1}{12}(1)(4)^3 + \frac{1}{12}(8)(1)^3 + \frac{1}{12}(1)(8)^3$$

$$I_{zz} = 48.67 \, in^4$$

Compute I_{xx} (Find Centroid)

$$\bar{z} = \frac{(8)(1)(1/2) + (8)(1)(5) + (4)(1)(9.5)}{8 + 8 + 4} = 4.10''$$

$$\therefore I_{xx} = \frac{1}{12}(8)(1)^3 + (8)(4.1-0.5)^2 + \frac{1}{12}(1)(8)^3 + (8)(5-4.1)^2$$
$$+ \frac{1}{12}(4)(1)^3 + (4)(9.5-4.1)^2 = 272 \, in^4$$

Compute P_{CR} for xy (Pinned-Pinned)

$$P_{CR} = \frac{\pi^2 EI}{L^2} = \frac{\pi^2 (29 \times 10^6 psi)(48.67 \, in^4)}{(144 \, in)^2} = 672 \, kips$$

Compute P_{CR} for zy (Fixed-free)

$$P_{CR} = \frac{\pi^2 EI}{4L^2} = \frac{\pi^2 (29 \times 10^6 psi)(272 \, in^4)}{4(144)^2} = 939 \, kips$$

USE 672 KIPS (MAX LOAD)

FBD

$$\tan\theta = 5/12 \quad \therefore \theta = \tan^{-1}(5/12) = 22.6°$$

$$\sum F_y = 0 \quad 672 - T\cos 22.6 = 0 \quad \therefore T = 728 \, kips$$

$$\sum F_x = 0 \quad F - T\sin\theta = 0$$
$$F = T\sin\theta = 728(\sin 22.6) = 280$$

$$\boxed{F_{MAX} = 280 \, KIPS}$$

EXAM 3, SET 2, PROBLEM 5

Orient the Axes SUCH THAT
$$\begin{cases} \theta_a = 0° & \varepsilon_a = -450\times10^{-6} \\ \theta_b = 120° & \varepsilon_b = -600\times10^{-6} \\ \theta_c = 240° & \varepsilon_c = -300\times10^{-6} \end{cases}$$

USING THE STRAIN TRANSFORMATION EQUATIONS:

$$\varepsilon_a = \varepsilon_x \cos^2\theta_a + \varepsilon_y \sin^2\theta_a + \gamma_{xy} \sin\theta_a \cos\theta_a$$
$$\varepsilon_b = \varepsilon_x \cos^2\theta_b + \varepsilon_y \sin^2\theta_b + \gamma_{xy} \sin\theta_b \cos\theta_b$$
$$\varepsilon_c = \varepsilon_x \cos^2\theta_c + \varepsilon_y \sin^2\theta_c + \gamma_{xy} \sin\theta_c \cos\theta_c$$

$$\therefore \; \varepsilon_a = -450\times10^{-6} = \varepsilon_x + 0 + 0$$
$$\varepsilon_b = -600\times10^{-6} = \tfrac{1}{4}\varepsilon_x + \tfrac{3}{4}\varepsilon_y - 0.433\,\gamma_{xy} \quad\left.\begin{array}{c} \\ \\ \end{array}\right\} \text{SOLVE } \varepsilon_x \; \varepsilon_y \; \gamma_{xy}$$
$$\varepsilon_c = -300\times10^{-6} = \tfrac{1}{4}\varepsilon_x + \tfrac{3}{4}\varepsilon_y + 0.433\,\gamma_{xy}$$

$$\varepsilon_x = -450\times10^{-6} \qquad \gamma_{xy} = 346.4\times10^{-6} \quad\therefore \varepsilon_{xy} = 173.2\times10^{-6}$$
$$\varepsilon_y = -450\times10^{-6}$$

$$\varepsilon_1, \varepsilon_2 = \frac{-450\times10^{-6} + -450\times10^{-6}}{2} \pm \sqrt{\left(\frac{-450\times10^{-6} - -450\times10^{-6}}{2}\right)^2 + (173.2\times10^{-6})^2}$$

$$= -450\times10^{-6} \pm 173.2\times10^{-6}$$

(a) $\therefore$ $\boxed{\varepsilon_1 = -277\times10^{-6} \quad \varepsilon_2 = -623\times10^{-6}}$

$$(\varepsilon_{xy})_{MAX} = \sqrt{\left(\frac{-450\times10^{-6} - -450\times10^{-6}}{2}\right)^2 + (173.2\times10^{-6})^2} = 173.2\times10^{-6}$$

(b) $\boxed{(\varepsilon_{xy})_{MAX} = 173.2\times10^{-6}}$

AVG NORMAL STRAIN $\boxed{-450\times10^{-6} = \varepsilon_{AVG}}$

EXAM 3, SET 2, PROBLEM 6

MULTIPLY LOADS BY 1.5 (FoS)

$18(15) = 2.7$ kips 2.25 kips $= 1.5(1.5)$

$8'$ $8'$ $10'$ $12'$

FBD BC
2.25
$10'$ $12'$ R_C
$$\Sigma M_B = 0 \quad R_C(22) - 2.25(10) = 0$$
$$\therefore R_C = 1.0227 \text{ kips}$$

FBD ABC
2.7 2.25
1.0227
$$\Sigma F_y = 0$$
$$0 = 1.0227 - 2.7 - 2.25 + R_A$$
$$\therefore R_A = 3.9273 \text{ kips}$$

$$\Sigma M_A = 0 \quad M_A - 2.7(8) - 2.25(26) + 1.0227(38) = 0 \;\therefore M_A = 41.2374 \text{ kip·ft}$$

3.9273 2.7 2.25 1.0227
41.2374 kip·ft $8'$ $8'$ $10'$ $12'$ kips

V (kip)
3.9273 3.9273
31.4184 22.0914 1.2273
1.0227

M (kip·ft)
1.0227 16.273
9.814
41.2374

$\therefore$ $\boxed{|V_{MAX}| = 3.9273 \text{ kips}}$
$\boxed{|M_{MAX}| = 41.2374 \text{ kip·ft}}$

$\sigma_{ALL} = 24$ ksi $\tau_{ALL} = 14$ ksi

$$\sigma_{ALL} = \frac{M_{MAX}}{S_{REQ'D}} \quad\therefore\; S_{REQ'D} = \frac{M_{MAX}}{\sigma_{ALL}} = \frac{(41,273)(12) \text{ in·lb}}{24,000 \text{ lb/in}^2}$$

$$\therefore S_{REQ'D} = 20.64 \text{ in}^3$$

FROM the Chart $\boxed{W\,12\times19}$

Check Shear
Area of Web: $\begin{cases} d = 12.2\text{ in} \\ t_w = 0.235\text{ in} \end{cases} A = 2.867\text{ in}^2$

$V_{MAX} = 3927.3 \text{ lb}$
$$\tau_{NET} = \frac{3927.3 \text{ lb}}{2.867\text{ in}^2} = 1,370 \text{ psi} < 14,000 \text{ psi}$$

$\therefore$ $\boxed{\text{USE } W\,12\times19}$

Properties of Geometric Shapes and Areas

Shape		$\bar{x}$	$\bar{y}$	Area	Length	I
Rectangular Area		$\dfrac{b}{2}$	$\dfrac{h}{2}$	bh		$I_x=\dfrac{1}{12}bh^3$ $I_y=\dfrac{1}{12}hb^3$
Triangular Area		$\dfrac{b}{3}$	$\dfrac{h}{3}$	$\dfrac{bh}{2}$		$I_x=\dfrac{1}{36}bh^3$
Quarter-circular Arc		$\dfrac{2r}{\pi}$	$\dfrac{2r}{\pi}$		$\dfrac{\pi r}{2}$	
Quarter-circular Area		$\dfrac{4r}{3\pi}$	$\dfrac{4r}{3\pi}$	$\dfrac{\pi r^2}{4}$		
Semicircular Arc		0	$\dfrac{2r}{\pi}$		πr	
Semicircular Area		0	$\dfrac{4r}{3\pi}$	$\dfrac{\pi r^2}{2}$		$I_x=\dfrac{\pi r^4}{8}-\dfrac{8r^4}{9\pi}$ $I_y=\dfrac{\pi r^4}{8}$
Semiparabolic Area		$\dfrac{3a}{8}$	$\dfrac{3h}{5}$	$\dfrac{2ah}{3}$		
Parabolic Area		0	$\dfrac{3h}{5}$	$\dfrac{4ah}{3}$		
Parabolic Spandrel		$\dfrac{3a}{4}$	$\dfrac{3h}{10}$	$\dfrac{ah}{3}$		
Circular Area		0	0	πr^2		$I_x=\dfrac{1}{4}\pi r^4$ $I_y=\dfrac{1}{4}\pi r^4$